MODULAR SERIES ON SOLID STATE DEVICES

VOLUME III

The Bipolar Junction Transistor

Second Edition

MODULAR SERIES
ON SOLID STATE DEVICES
Gerold W. Neudeck and Robert F. Pierret, Editors

VOLUME III

The Bipolar Junction Transistor
Second Edition

GEROLD W. NEUDECK
Purdue University

ADDISON-WESLEY PUBLISHING COMPANY
READING, MASSACHUSETTS
MENLO PARK, CALIFORNIA • NEW YORK
DON MILLS, ONTARIO • WOKINGHAM, ENGLAND
AMSTERDAM • BONN • SYDNEY • SINGAPORE
TOKYO • MADRID • SAN JUAN

This book is in the
Addison-Wesley Modular Series on Solid State Devices

Library of Congress Cataloging-in-Publication Data

Neudeck, Gerold W.
    The bipolar junction transistor / by Gerold W. Neudeck. — 2nd ed.
      p. cm. — (Modular series on solid state devices; v. 3)
    Bibliography: p.
    Includes index.
    ISBN 0-201-12297-9
    1. Bipolar transistors. 2. Junction transistors. I. Title.
II. Series.
TK7871.96.B55N48 1989
621.3815'282 — dc19   88-22601
                                                                    CIP

Reprinted with corrections September, 1989

11 12 13 14 15 16 17 18 19 20 - CRS - 98 97 96 95

Foreword

Solid state devices have attained a level of sophistication and economic importance far beyond the highest expectations of their inventors. By continually offering better-performing devices at lower cost per unit, the electronics industry has penetrated markets never before addressed. A prerequisite to sustaining this growth initiative is an enhanced understanding of the internal workings of solid state devices by modern electronic circuit and systems designers. This is essential because system, circuit, and IC layout design procedures are being merged into a single function. Considering the present and projected needs, we have established this series of books (the Modular Series) to provide a strong intuitive and analytical foundation for dealing with solid state devices.

Volumes I through IV of the Modular Series are written for juniors, seniors, and possibly first-year graduate students who have had at least an introductory exposure to electric field theory. Emphasis is placed on developing a fundamental understanding of the internal workings of the most basic solid state device structures. With some deletions, the material in the first four volumes is used in a one-semester, three-credit-hour, junior-senior course in electrical engineering at Purdue University. The material in each volume is specifically designed to be presented in ten to twelve 50-minute lectures.

The volumes of the series are relatively independent of each other, with certain necessary formulas repeated and referenced between volumes. This flexibility enables one to use the volumes sequentially or in selected parts, either as the text for a complete course or as supplemental material. It is also hoped that students, practicing engineers, and scientists will find the series useful for individual instruction, whether it be for reference, review, or home study.

A number of the standard texts on devices have been written like encyclopedias, packed with information, but with little thought as to how the student learns or reasons. Texts that are encyclopedic in nature are often difficult for students to read and may even present barriers to understanding. It is hoped that by breaking the material into smaller units of information, and by writing *for students,* we have constructed volumes which are truly readable and comprehensible. We have also sought to strike a healthy balance between the presentation of basic concepts and practical information.

The problems presented at the end of each chapter constitute an important component of the learning program. Some of the problems extend the theory presented in the text or are designed to reinforce topics of prime importance. Others are numerical problems which provide the reader with an intuitive feel for the typical sizes of key parameters. When approximations are stated or assumed, the student will then have confidence that cited quantities are indeed orders of magnitude smaller than others. The end-of-chapter problems range in difficulty from very simple to quite challenging. In the second edition we have added a new feature—worked problems or *exercises*. The exercises are collected in Appendix A and are referenced at the appropriate points within the text. The exercises are similar in nature to the end-of-chapter problems. Finally, Appendix B contains sets of volume-review problems and answers. These sets contain short-answer, testlike questions which could serve as a review or as a means of self-evaluation.

Reiterating, the emphasis in the first four volumes is on developing a keen understanding of the internal workings of the most basic solid state device structures. However, it is our hope that these volumes will also help (and perhaps motivate) readers to extend thier knowledge—to learn about the many more devices already in use and even to seek information about those presently under investigation in research laboratories.

Prof. Gerold W. Neudeck
Prof. Robert F. Pierret
Purdue University
School of Electrical Engineering
W. Lafayette, IN 47907

Contents

Introduction

This volume concentrates on the bipolar junction transistor (BJT), presenting both qualitative and quantitative descriptions of the device. The two basic forms of the bipolar transistor are the *npn* and *pnp*, named from the three layers of semiconductor used to construct the device. Both forms are in general use as individual discrete devices and in integrated circuits. Because of its generally higher gain and faster switching, the *npn* version is preferred in many circuit designs.

To make the bipolar presentation efficient, many of the concepts and derivations developed in Volume II for the junction diode are applied directly to the junction transistor. However, the bipolar transistor differs from the diode in that the transistor is capable of current gain, voltage gain, and power gain. It is an active device whereas the diode, like a resistor, is a passive device. We therefore emphasize the ability of the transistor to achieve gain.

The *npn* and *pnp* bipolar transistors are *complementary*. This means that the devices have an interchange of material types (*n* to *p* and *p* to *n*) and have opposite polarities for all voltages and currents. We will use this idea of complements, rather than repeat the physical descriptions, currents, and voltages for both the *npn* and *pnp* devices. The following chapters use the *pnp* device in developing the concepts and equations describing the bipolar transistor, because the directions of the currents and carrier fluxes are simpler to describe. Since the *npn* is the complement of the *pnp*, one needs only to change all current and voltage polarities to describe the *npn*.

Chapter 1 establishes the reference directions for the currents and voltages as well as a qualitative description of bipolar transistor action and fabrication. We also define the regions of operation and the carrier fluxes for those regions. Chapter 2 outlines a "game plan" for the solution of carrier concentrations in the base and presents the equations necessary to obtain the terminal currents as a function of the terminal voltages. An ideal bipolar transistor is defined and the terminal currents are derived. Chapter 2 also discusses the regions of operation of the ideal *pnp* and presents the common emitter input and output V–I characteristics. The dc values for alpha, beta, emitter injection efficiency, I_{CB0}, and I_{CE0} are derived and discussed qualitatively. The Ebers–Moll equations for the ideal device are derived and are noted to be valid for all four regions of operation. These equations embody the nonlinear dc model for bipolar transistor devices, and serve a role similar to that of the ideal diode equation for the diode.

The deviations of a real device from the ideal transistor are the subject of Chapter 3. Base recombination is first presented and then simplified for the quasi-ideal BJT. Base width changes, avalanche breakdown, reach-through, generation–recombination, and graded base effects are discussed. Chapter 4 derives the small-signal, low-frequency, hybrid-pi model for the ideal transistor. Nonideal effects are then included to complete the low-frequency model. The depletion and diffusion capacitances for the device are summarized and added to obtain a high-frequency signal model. Finally, Chapter 5 presents the charge control model and derives the storage time for a saturated transistor that is switched to cutoff with a base current step. The turn-on transient is also derived.

1 / Introduction to Bipolar Junction Transistors

1.1 TERMINOLOGY AND SYMBOLS

The junction transistor is, by definition, a semiconductor device containing three adjoining, alternately doped regions in which the middle region is very narrow compared with the minority carrier diffusion length for that region. As shown in Fig. 1.1(a), the external world contact to the narrow central region is known as the *base*. Contacts to the outer regions are labeled the *emitter* and *collector*. The emitter and collector designations arise from the functions performed by these regions in the operation of the device; it would appear from Fig. 1.1(a) that these two regions might be interchangeable. However, in modern-day practical devices, the emitter region is usually much more heavily doped than the collector and the terminals cannot be interchanged without changing the device's characteristics.

Figure 1.1(b) illustrates the circuit symbol for the *pnp* junction transistor while simultaneously defining the pertinent current and voltage polarities. Although "+" and "−" signs are shown in this figure to define voltage polarities, they are actually redundant, because the double subscript on the voltage symbol likewise denotes the voltage polarity. The first subscript indicates the assumed reference polarity of "+." For ex-

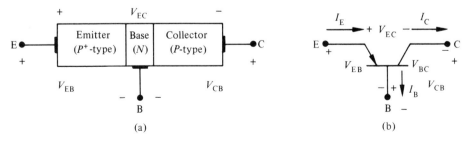

Fig. 1.1 *pnp* bipolar transistor: (a) semiconductor types; (b) circuit symbol with voltage and current polarities.

ample, V_{EB} assumes the "E" to have the "+" sign and "B" the "−" sign. Note that as a consequence of Kirchhoff's circuit laws there are only two independent voltages and two independent currents. If two currents or voltages are known, then the third is also known.

Throughout this volume, current reference directions were chosen to coincide with the physical current flow in the active region for each device rather than the IEEE standard notation of all currents for all devices entering each terminal. With the current reference directions as chosen, all currents in the active region are positive quantities; this makes conceptual arguments much easier for the reader. It also provides a more intuitive base for understanding the device physics of carrier flow.

Figure 1.2(a) illustrates the *complement* of the *pnp*, the *npn* transistor. By "complement" we mean the interchange of *p* for *n* and *n* for *p*. The current and voltage reference directions are displayed in Fig. 1.2(b) for the *npn*. When we compare Figs. 1.1(b) and 1.2(b), it is clear that to obtain the complement of the *pnp*, the *npn*, all the current and voltage polarities are reversed. If you understand the *pnp*, then you need only reverse polarities and conduction types to describe the *npn*. This volume will concentrate on the p^+np, because it follows more directly the format of the equations and polarities developed in Volume II for the junction diode. However, the n^+pn is more commonly used in most *IC* designs.

The *pnp* transistor of Fig. 1.1(a) can be thought of as two very closely spaced p-n junctions. One junction is formed between the emitter and base, and the other between the collector and base regions. The *n*-type base region is typically less than 1 μm in width (much less than a minority carrier diffusion length). Because of the close proximity of the two junctions, they interact with each other, and thus the transistor is capable of current and voltage gain. A more detailed discussion of their interaction and gain is presented in Chapter 2.

The bipolar transistor has four *regions of operation* or dc bias. By "regions of operation" we mean the voltage polarities on the collector-to-base junction and the emitter-to-base junction. For example, the junction diode had two regions of operation,

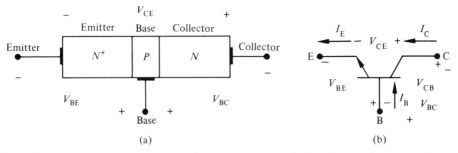

Fig. 1.2 *npn* bipolar transistor: (a) semiconductor types; and (b) voltage and current reference polarities.

forward and reverse bias, depending on the junction voltage polarity. The most common region of operation for the bipolar transistor is the *active region* defined as having the E–B junction forward biased and the C–B junction reverse biased. For the p^+np this means that the E–B has a "+" to "−" polarity and the C–B has a "−" to "+" polarity. Almost all linear signal amplifiers, such as operational amplifiers, have their bipolar transistors biased in the active region because they have their largest signal gain and smallest signal distortion in this region.

The *saturation region* is defined as having both the E–B junction and the C–B junction forward biased. For the *pnp* this means positive V_{EB} and V_{CB} voltages. In logic circuits and transistor switches, this represents the region of operation where $|V_{CE}|$ is small and $|I_C|$ is large; that is, the device acts like an "on" switch. A closed (or "on") switch has little or no voltage drop across it even with large current flow. In a logic circuit we call this the zero, or "low," logic level.

The *cutoff region* is defined as having both junctions reverse biased. A negative voltage polarity of V_{EB} and V_{CB} is necessary for the *pnp* transistor. Typically this represents the "off" state for the transistor as a switch, or the "high" logic level in digital circuits. When "off," the transistor is similar to an open switch in that $|I_C|$ is nearly zero and $|V_{CE}|$ is large.

The fourth region of operation is the *inverted region*, which is sometimes called the *inverted active region*. For inverted active operation, the E–B is reverse biased and the C–B is forward biased. One might think of the inverted active case as one in which the collector acts like an emitter and the emitter like a collector; that is, the device is used backwards. The most common use of this region of operation is in digital logic circuits such as TTL logic where signal gain is not an objective.

Figure 1.3(a) illustrates the V_{EB} and V_{CB} voltages for the four *pnp* regions of operation. Note that the "complement" of Fig. 1.3(a), Fig. 1.3(c), for the n^+pn has V_{BE} and V_{BC} for the axes. Figure 1.3(b) illustrates the regions of operation on the output V–I characteristic for the p^+np while Fig. 1.3(d) is for the n^+pn device. Chapter 2 discusses each of these regions in much greater detail.

In circuit applications the transistor typically functions with a common terminal between the input and output, either dc common or signal ground common. Because the transistor has only three leads, there are three possible amplifier types. The designations are *common base*, *common emitter*, and *common collector;* these names indicate the lead common to both the input and output circuits as illustrated in Figs. 1.4(a), (b), and (c), respectively, for the *pnp*. The most often used of the three amplifier types is the common emitter. In describing the amplifier and its V–I characteristics, the current in the "common" lead is the variable eliminated. For example, the common emitter amplifier has output variables v_{EC} and i_C. Its input variables are v_{EB} and i_B. Again note that if two of the voltages (or currents) are known, the third is also known from Kirchhoff's Laws. Therefore, to describe the common emitter by its output and input V–I characteristics implies that you can calculate the common base and common collector characteristics. The *npn* is the complement by reversing all current and voltage polarities.

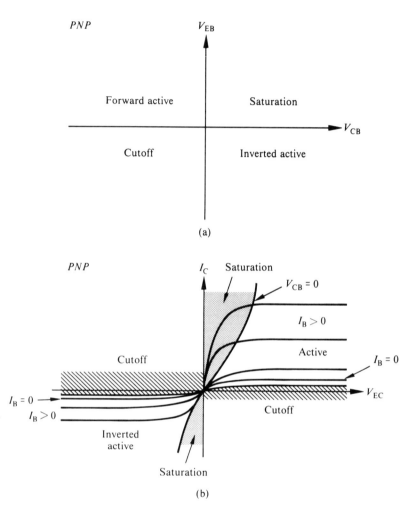

Fig. 1.3 Regions of operation: (a) p^+np junction polarities; (b) p^+np output characteristic; (c) n^+pn junction polarities; (d) n^+pn output characteristic.

SEE EXERCISE 1.1–APPENDIX A

1.2 QUALITATIVE OPERATION OF THE ACTIVE REGION

The p^+np bipolar junction transistor (BJT) is first discussed since it will follow much of the discussion and use some of the results of the p^+n junction of Volume II. Since the n^+pn BJT is used extensively in many circuit applications, it is also presented. The

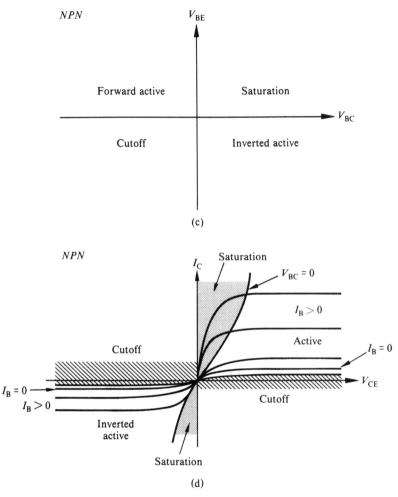

Fig. 1.3 (continued)

purpose of this section is to develop *how* the BJT works from a qualitative point of view before jumping into the mathematical descriptions of Chapter 2.

1.2.1 The p^+np Transistor

Before one can understand how a transistor works it is necessary to establish a few basic facts. To do this let us consider a *pnp* transistor in thermal equilibrium. Figure 1.5(a) illustrates the energy band diagram for uniformly doped regions where the emitter is doped heavier *p*-type than is the collector. The magnitude of the *n*-type base doping is less than the emitter, but greater than the collector. Note that this figure is

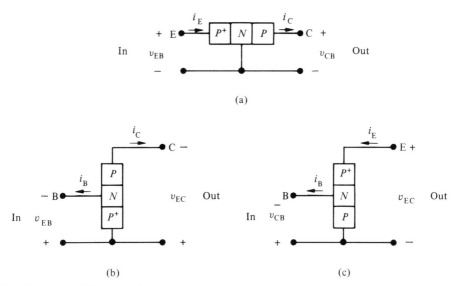

Fig. 1.4   Amplifier types for a *pnp*: (a) common base; (b) common emitter; (c) common collector.

just an extension of the thermal equilibrium case for the *p-n* junction, as applied to a p^+-*n* and an *n-p* junction closely spaced. All of the physical arguments presented in Volume II, Chapter 2, are applicable to the E–B and B–C junctions. The built-in potential (V_{bi}) charge density, electric fields, etc., as discussed, are still valid for each junction. Figures 1.5(b) and (c) illustrate the charge density and electric fields in the two depletion regions. The potential, $V(x)$, plot is the negative image of the energy band diagram and is not shown. Note in Fig. 1.5(a) that the depletion regions are labeled W_E and W_C for the emitter-base and collector-base junctions. Also note that the base bulk region is now labeled W and *should not be confused with the diode depletion region.*

At thermal equilibrium there is no net current flow; hence, as was the case for the junction diode, all the drift and diffusion currents cancel each other. Said another way, all the drift components are equal and opposite to the diffusion components for each carrier type in each junction. If ideal junctions are assumed, then all the depletion widths, electric fields, etc., are calculated from the formulas presented for the ideal diode in Chapter 2 of Volume II.

The *pnp* device biased in the active region requires the emitter to have a higher potential, and the collector a lower potential, than that of the base. Figure 1.6(a) shows the energy band diagram for thermal equilibrium and for active region operation. In this energy band diagram we have let the base stay fixed and have lowered the emitter due to the positive voltage applied to that bulk region. When the E–B junction is forward biased, note that the barrier for holes entering the *n*-type base region from the

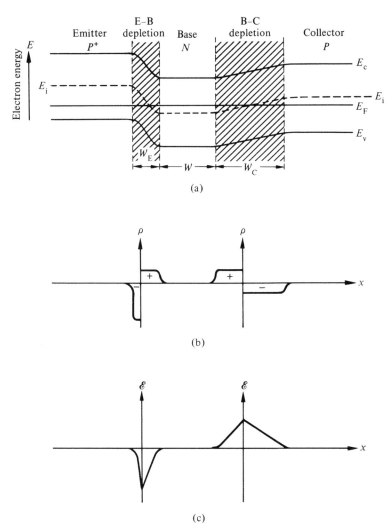

Fig. 1.5 p^+np in thermal equilibrium: (a) energy band diagram; (b) charge density; (c) electric field.

emitter is lowered* as was the case in a forward biased diode; holes are allowed to be injected from the emitter into the base. Also note that the barrier for electrons in the base is lowered and electrons are injected from the base into the emitter. The combined effect is that the forward biased E–B junction creates a positive emitter current. As was the case for the diode, the diffusion currents for electrons and holes have increased exponentially with an increase in V_{EB}.

*Remember that holes like to "float," and electrons to sink, in an electron energy band diagram.

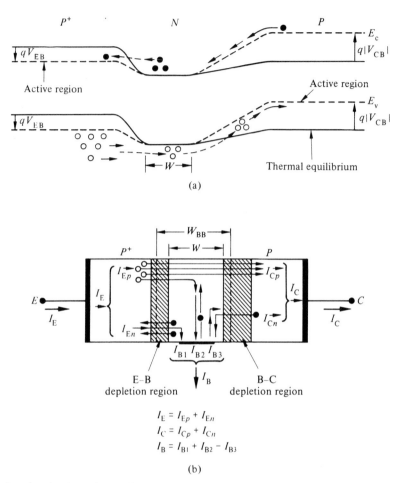

Fig. 1.6 p^+np in the active region: (a) energy band diagram (——thermal equilibrium, - - - active region); (b) current components and carrier flux.

The collector-base junction is reverse biased for active region operation and the right-hand side (the collector) of Fig. 1.6(a) is raised with respect to the n-type base region. This effectively raises the barrier for electrons in the base that may want to travel to the collector. Similarly, the hole barrier from collector to base is increased. However, note that the holes injected from (coming from) the emitter into the very narrow (<1 μm) base easily diffuse through the base and find a potential hill to slide down* into the collector. The reader may now get an inkling of the terminology of emitter (injector of carriers) and collector (collector of those injected elsewhere). If almost all the injected holes from the emitter traverse the base with $<1\%$ recombining in

---

*Remember that holes like to "float," and electrons to sink, in an electron energy band diagram.

the n-base region, then the collector hole current is slightly less than the emitter hole current. The electron component of the collector current arises from the thermally generated minority electrons at the edge of the p-type collector that fall down the potential hill into the base.

For any "interaction" to take place between the two transistor junctions, they must be spaced such that the base width is much less than the minority (holes in the base) carrier diffusion length. If this is not the case, then the pnp structure is nothing more than two back-to-back diodes; that is, all the holes injected from the emitter recombine in the base and never reach the reverse biased C–B junction.

Looking inside the device, what do we see as far as the electrostatics are concerned? First, around both the emitter-base and collector-base metallurgical junctions there is a depletion region. The E–B space charge region will be smaller than in equilibrium because this junction is forward biased. The C–B junction depletion region will be wider than in equilibrium because this junction is reverse biased. Hence, we can visualize the situation to be something like that shown in Fig. 1.6(b). Note that part of the base, the width W, is quasi-neutral; that is, it contains essentially no electric field. It should be emphasized that the quasi-neutral base width W is *not* equal to the metallurgical base width, W_{BB}. Specifically, W is the width of the metallurgical base less the regions of the base that are depleted.

In the preceding discussion we concentrated on the hole current flowing from the emitter to the collector to establish several important facts. Further qualitative information can be gained by examining the base current components in the system. This will be done with the aid of Fig. 1.6(b), which diagrams all the various major* current components (other than recombination–generation currents in the depletion region, which we ignore for the time being). The three active region base current components shown in this figure have been labeled I_{B1}, I_{B2}, and I_{B3}. Together they make up the total current going out of the base as follows: (1) I_{B1} corresponds to the current arising from electrons being back injected (diffusing) across the forward biased E–B junction from the base into the emitter. (2) I_{B2} corresponds to those electrons that must enter the base to replace electrons used up in recombining with holes injected from the emitter. Only a very small number of injected holes will recombine, with most passing on through the base into the collector. (3) I_{B3} is that part of the collector current due to thermally generated electrons in the collector that are within one diffusion length of the C–B junction edge and that fall down the potential hill from the collector into the base.

Practical transistors are fabricated such that the emitter doping is much greater than the base doping, which in turn is made much greater than the collector doping. What effect does this have on the size of the current in the transistor? Because the emitter doping is much greater than that of the base, the hole current across the E–B junction (I_{Ep}) will be much greater than the electron current ($I_{B1} = I_{En}$), as was the case in the p^+-n diode. Hence, I_E is approximately equal to the hole current injected into the base;

---

*Thermally generated elecrons in the emitter are neglected because of the large doping, hence the very small number of carriers generated. Holes generated in the base also are neglected since W is small and therefore not many carriers are generated.

that is, $I_{En} \ll I_{Ep}$ and $I_E \cong I_{Ep}$. Because W is made $\ll L_p$, very few holes will be lost as they traverse the base, and I_{B2} will also be much less than I_E. Finally, because I_{B3} corresponds to a reverse bias saturation current of the C–B junction, it too will be very small and I_C will be about equal to the hole current passing through the base. Adding these facts together, one concludes that for practical transistors under normal active region biasing conditions:

1. The injected hole current from E–B is $I_{Ep} \cong I_E$ with I_E being slightly larger than I_C.
2. $I_B \ll I_C$ or I_E.

The large current gain capability of the bipolar transistor, for a common emitter device, results from the fact that a small base current provides electrons for recombination with holes coming from the emitter and for back injection into the emitter. The current gain I_C/I_B is large because a p^+n junction (E–B) needs only a small electron current to provide a large hole current. Stated somewhat differently, a small base current forces the E–B to become forward biased and inject large numbers of holes which travel through the base to the collector.

1.2.2   The n^+pn Transistor

Consider the energy band diagram for an npn transistor in thermal equilibrium, similar to Fig. 1.6(a), as shown in Fig. 1.7(a). This figure illustrates the energy band diagram for uniformly doped regions where the emitter is doped heavier n-type than is the collector. The magnitude of the p-type base doping is less than that of the emitter, but greater than that of the collector. Note that this figure is just an extension of the thermal equilibrium case for the p-n junction as applied to an n^+-p and a p-n junction that are closely spaced. At thermal equilibrium there is no net current flow; hence the drift and diffusion currents for each carrier type cancel each other for each junction.

Figure 1.7(a) shows the energy band diagram for thermal equilibrium and for active region operation. The npn device biased in the active region requires the emitter to have an energy higher than that of the base due to the negative potential with respect to the base. Note that the barrier for electrons entering the p-base region, from the emitter, is lowered and electrons are allowed to be injected from the emitter into the base. Also note that the barrier for holes in the base is lowered and holes are injected from the base into the emitter. The combined effect is that the forward biased B–E junction creates the positive current leaving the emitter. Remember that conventional current flow is opposite to electron flow.

The collector-base junction is reverse biased for active region operation and the right-hand side (the collector) of Fig. 1.7(a) is lowered with respect to the p-base region. This effectively raises the barrier for holes in the base that may want to travel to the collector. Similarly, the electron barrier from the collector to the base has increased. However, note that the electrons injected from the emitter into the very narrow (<1 μm) base easily diffuse through the base and find a potential hill to slide down into the collector. If almost all the injected electrons from the emitter traverse the

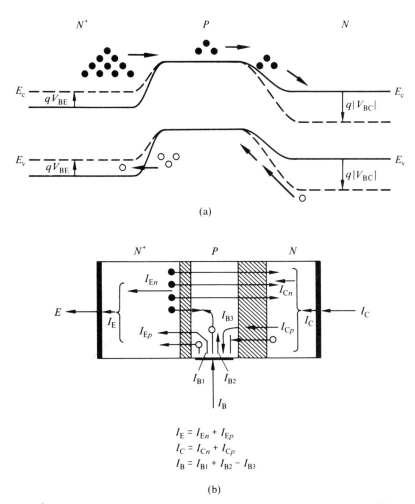

Fig. 1.7 n^+pn in the active region: (a) energy band diagram (———thermal equilibrium, - - - active region); (b) current components and carrier flux.

base, then the collector electron current is slightly less than the emitter electron current. The hole component of the collector current arises from the thermally generated minority electrons at the edge of the n-type collector that fall "down" the potential hill for holes into the base.

Further qualitative information can be gained by examining the base current components in the system. This will be done with the aid of Fig. 1.7(b), which diagrams all the various current components (other than recombination–generation currents in the depletion region). The three base current components shown in this figure have been labeled I_{B1}, I_{B2}, and I_{B3}. Together they make up the total base current going into the base, as follows. (1) The current I_{B1} corresponds to the current arising from holes being back injected across the forward biased B–E junction from the base into the emitter. (2) I_{B2} corresponds to those holes that must enter the base to replace holes used up in recombining with electrons injected from the emitter. Only a very small number of injected electrons will recombine, with most passing on through the base into the collector. (3) Current I_{B3} is that part of the collector current due to thermally generated holes in the collector that are within one diffusion length of the B–C junction edge and that fall down the potential hill from the collector into the base.

Most bipolar transistors are fabricated such that the emitter doping is much greater than the base doping, which in turn is made much greater than the collector doping. Because the emitter doping is much greater than that of the base, the electron current across the B–E junction (I_{En}) will be much greater than the hole current ($I_{Ep} = I_{B1}$). Hence, I_E is approximately equal to the electron current injected into the base. Because W is made $\ll L_N$, very few electrons will be lost as they traverse the base and I_{B2} will also be much less than I_E. Finally, because I_{B3} corresponds to a reverse bias saturation current, it too will be very small. Hence I_C will be about equal to the electron current passing through the base.

1.3 FABRICATION

The fabrication of *pnp* and *npn* bipolar transistors is similar to that of *p-n* junction diodes as described in Volume II, Chapter 1 and discussed extensively in Volume V of this series. The major difference is that two *p-n* junctions must be formed for a *pnp* device — a p^+-*n* junction for the E–B and an *n-p* junction for the C–B, with the base *n*-region being 1 μm or less in width.

Figures 1.8(a) and (b) illustrate a typical discrete, double-diffused p^+np transistor and an integrated circuit n^+pn transistor, respectively. The discrete *pnp* is formed by starting with a p^+-type (heavily doped) substrate and growing a high-resistivity (low-doping) *p*-type epitaxial layer of silicon on the surface. The *p*-epitaxial layer is typically about 5 to 10 μm thick. The *n*-base region is thermally diffused through an oxide window into the *p*-type epitaxial layer. Now a p^+ (very heavily doped) emitter is diffused into the *n*-base region. The geometrical arrangements of Fig. 1.8 are not to scale. To get a better idea of the size, stretch the figure 50 times in the horizontal direction, leaving the vertical dimensions fixed. The metal is an aluminum-silicon alloy and makes an ohmic contact with the three regions.

The integrated circuit *npn* transistor of Fig. 1.8(b) is fabricated starting with a high-resistivity *p*-type substrate. A small area of n^+ is diffused into the *p*-substrate and is called a buried layer or subcollector. The primary function of the buried layer is to provide a low-resistance path for the collector current to reach the collector contact, which now must be at the top surface. After the n^+ diffusion, a lightly doped *n*-type epitaxial layer is grown over the n^+ and *p*-substrate. The epitaxial *n*-region becomes the low-doped collector.

Isolation between resistors and transistors (components) in a bipolar integrated circuit is achieved with the use of a reverse biased diode, and is therefore called *diode*

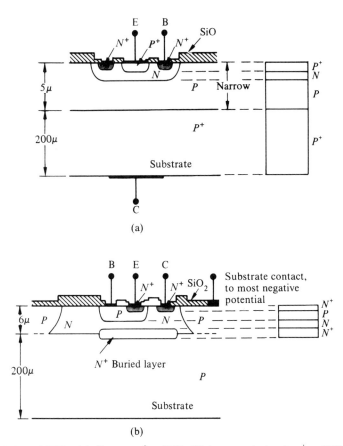

Fig. 1.8 Fabricated BJTs: (a) discrete p^+np BJT; (b) integrated circuit n^+pn BJT.

isolation. The *p*-regions are diffused through an oxide mask completely through the *n*-epitaxial layer to make contact with the *p*-substrate. By placing the *p*-substrate at the most negative potential in the circuit, the *n*-epitaxial region is at some higher potential and therefore the *n*-epitaxial region is surrounded by a reverse biased *p-n* junction. In more advanced devices the vertical junctions have been replaced by silicon dioxide isolation.

The *p*-type base is diffused into the *n*-epitaxial region followed by the n^+ emitter and collector contact region impurity diffusions. The collector contact at the surface is necessary in order to interconnect components on the surface, while the n^+n collector contact is required because aluminum on n^+ forms an ohmic contact as discussed in Volume II, Chapter 7. Aluminum on lightly doped *n*-silicon can form a rectifying contact, a Schottky diode. Remember that aluminum is a *p*-type dopant, so its contact to the *p*-base forms an ohmic contact.

In the remaining chapters we assume a one-dimensional, bipolar analysis; that is, the major current flow is in only one direction. Because of the very narrow vertical dimension and little lateral current flow (except to the base contact), the assumption yields very good first-order models. In particular, it gives equations that can be solved without reverting to a supercomputer for numerical integrations.

1.4 CIRCUIT DEFINITIONS

The circuit variables necessary for the description of the bipolar transistor in the active region are presented in this section. Figure 1.9(a) illustrates the major current components of the emitter and collector currents for the *pnp* BJT. The holes injected from the emitter to the base make up the current component I_{Ep}, and the electrons back injected from the base form I_{En}. Those holes injected from the emitter that reach the collector junction constitute I_{Cp}. The current I_{Cn} is a result of the thermally generated electrons near the C–B junction that drift into the base. Equations (1.1) through (1.3) are the terminal currents written in terms of the active region current components.

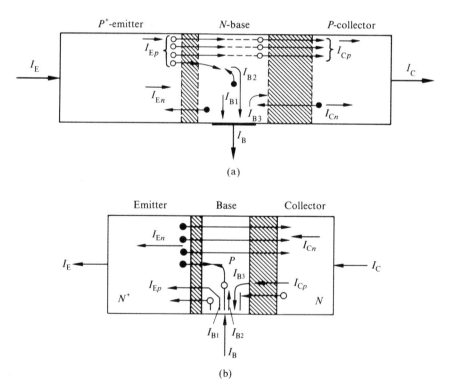

(a)

(b)

Fig. 1.9 Carrier flux and current components for (a) active region operation p^+np and (b) active region operation n^+pn.

$$I_E = I_{Ep} + I_{En} \tag{1.1}$$

$$I_C = I_{Cp} + I_{Cn} \tag{1.2}$$

$$I_B = I_E - I_C = I_{B1} + I_{B2} - I_{B3} \tag{1.3}$$

The *base transport factor* for the p^+np is defined as the ratio of the hole current diffusing into the collector to the hole current injected at the E–B junction and is expressed as

$$\boxed{\alpha_T = I_{Cp}/I_{Ep}} \qquad (pnp) \tag{1.4}$$

Ideally, α_T would be unity; however, with recombination of holes in the base, α_T is slightly less than unity. In well fabricated narrow base devices, $\alpha_T \rightarrow 1.0$.

Another performance parameter is the *emitter injection efficiency* (γ) which measures the ratio of the injected hole current to total emitter current:

$$\boxed{\gamma = \frac{I_{Ep}}{I_E} = \frac{I_{Ep}}{I_{En} + I_{Ep}}} \qquad (pnp) \tag{1.5}$$

Note that $\gamma \rightarrow 1$ if $I_{En} \rightarrow 0$; that is, as the emitter is more heavily doped, I_{En} becomes a smaller percentage of I_E (similar to the p^+-n diode current components). For larger-gain transistors, γ is made as close to unity as possible.

The ratio of I_C/I_E in the active region is defined as the *dc alpha* (α_{dc}), and from Eqs. (1.1) and (1.2),

$$\alpha_{dc} = \frac{I_C}{I_E} = \frac{I_{Cp} + I_{Cn}}{I_{Ep} + I_{En}} \tag{1.6}$$

For almost any degree of E–B forward bias and C–B reverse bias, $I_{Cp} \gg I_{Cn}$ and therefore Eq. (1.6) is approximated by

$$\boxed{\alpha_{dc} = \frac{I_{Cp}}{I_{Ep} + I_{En}}} \tag{1.7}$$

If I_{Cp} and I_{Ep} are factored out of Eq. (1.7) and the definition of α_T is applied,

$$\alpha_{dc} = \frac{I_{Cp}}{I_{Ep}}\left[\frac{1}{1 + I_{En}/I_{Ep}}\right] = \alpha_T\left[\frac{I_{Ep}}{I_{Ep} + I_{En}}\right] \tag{1.8}$$

with the C–B current flowing with the emitter open circuited ($I_E = 0$), similar to the reverse saturation current of the n^+-p diode. The collector current is then obtained from Eq. (1.12) as

$$\boxed{\alpha_{dc} = \gamma\alpha_T} \tag{1.9}$$

Ideally, as γ and α_T approach unity, so will α_{dc}. What is good for γ and/or α_T is good for α_{dc}.

Beta (β_{dc}) is another performance parameter for the active region and is defined by

$$\beta_{dc} = \frac{I_C}{I_B} = \frac{I_C}{I_E - I_C} \tag{1.10}$$

Factoring I_E from the denominator of Eq. (1.10) yields

$$\beta_{dc} = \frac{I_C}{I_E(1 - I_C/I_E)} = \frac{I_C/I_E}{1 - I_C/I_E}$$

From Eq. 1.6 (the definition of α_{dc}),

$$\boxed{\beta_{dc} = \frac{\alpha_{dc}}{1 - \alpha_{dc}}} \tag{1.11}$$

Note that as $\alpha_{dc} \to 1$, $\beta_{dc} \to$ infinity. Since I_C/I_B is a current gain, we see the desirability for γ and α_T (therefore α_{dc}) to be as close to unity as possible.

The common base active region is most easily described by expanding Eq. (1.2) by using Eqs. (1.4) and (1.5):

$$I_C = I_{Cp} + I_{Cn} = \alpha_T I_{Ep} + I_{Cn} = \gamma\alpha_T\frac{I_{Ep}}{\gamma} + I_{Cn} = \alpha_{dc}I_E + I_{Cn} \tag{1.12}$$

The collector reverse saturation current, I_{Cn}, is defined as

$$I_{Cn} \cong I_{BCO} \tag{1.13}$$

with the C–B current flowing with the emitter open circuited ($I_E = 0$), similar to the reverse saturation current of the n^+-p diode. The collector current is then obtained from Eq. (1.12) as

$$I_C = \alpha_{dc} I_E + I_{BC0} \tag{1.14}$$

The common emitter, active region, collector current can be obtained by substituting Eq. (1.3) into Eq. (1.14):

$$I_C = \alpha_{dc}(I_B + I_C) + I_{BC0} \tag{1.15}$$

Solving for I_C,

$$I_C(1 - \alpha_{dc}) = \alpha_{dc} I_B + I_{BC0} \tag{1.16}$$

$$I_C = \frac{\alpha_{dc}}{(1 - \alpha_{dc})} I_B + \frac{I_{BC0}}{(1 - \alpha_{dc})} = \beta_{dc} I_B + I_{BC0}(\beta_{dc} + 1) \tag{1.17}$$

If the E–C reverse saturation current (I_{EC0}) is defined as the case of $I_B = 0$, then, from Eq. (1.17),

$$I_{EC0} = I_{BC0}(\beta_{dc} + 1) \tag{1.18}$$

and

$$I_C = \beta_{dc} I_B + I_{EC0} \tag{1.19}$$

We have now developed a set of equations for the active region operation for the *pnp* bipolar transistor. A similar set of equations is easily developed for the *npn* transistor using Fig. 1.9(b). For example, the equations

$$\alpha_T = \frac{I_{Cn}}{I_{En}} \qquad (npn) \tag{1.20}$$

$$\gamma = \frac{I_{En}}{I_{En} + I_{Ep}} \qquad (npn) \tag{1.21}$$

$$\alpha_{dc} = \frac{I_{Cn}}{I_{En} + I_{Ep}} = \frac{I_{Cn}}{I_E} \qquad (npn) \qquad (1.22)$$

can be obtained from the basic definitions. Note that Eqs. (1.9), (1.11), (1.14), (1.18), and (1.19) apply as they are to the *npn* device with only a change in I_{BC0} to I_{CB0} and I_{EC0} to I_{CE0}.

The task of Chapter 2 will be to derive the quantitative equations for I_C, I_E, and I_B in terms of the material parameters such as N_A, N_D, τ_p, etc., and in terms of the E–B and C–B voltages.

SEE EXERCISE 1.2–APPENDIX A

1.5 SUMMARY

The regions of operation for the *pnp* and the *npn* were defined in terms of the applied voltage conditions on the E–B and C–B junctions. Active region operation (largest signal gain) has the E–B junction forward biased and the C–B junction reverse biased. Saturation region operation occurs when both junctions are forward biased. Cutoff occurs when both junctions are reverse biased.

A qualitative description of how the BJT conducts carriers was investigated from the energy band and carrier flux viewpoints. From the energy band, net carrier flow under applied bias in the active region was related to the increase, or decrease, of the diffusion current components. When forward biased, the diffusion currents increase exponentially with the number of carriers with energies larger than the potential barrier. Under reverse bias they decrease with applied voltage. An alternate viewpoint that contains the same information was conveyed by the carrier flux diagrams illustrating the various current components. The base current was made small by doping the emitter larger than the base and by making the base small with respect to a minority carrier diffusion length.

Circuit definitions for α_{dc}, γ, α_T, and β_{dc} under active region operation were defined in terms of the current components. These definitions were then used to relate the collector, base, and emitter currents to each other. Remember that these are for active region operation only.

PROBLEMS

1.1 Determine the regions of operation for the bipolar junction transistor (BJT) if the polarities are as shown for the

(a) *pnp*

Region	V_{EB}	V_{CB}
	+	+
	−	−
	+	−
	−	+

(b) *npn*

Region	V_{BE}	V_{BC}
	+	+
	−	−
	+	−
	−	+

1.2 A silicon p^+np BJT is at thermal equilibrium. Let $n_i = 10^{+10}/cm^3$ and $kT = 0.026$ eV. If the doping densities are

$$N_{AE} = 5 \times 10^{17}, \qquad N_{DB} = 10^{15}, \qquad N_{AC} = 10^{14}/cm^3$$

(a) Sketch the energy band diagram and show energy level positions in units of kT.

(b) Sketch the charge density and calculate the maximum electric fields.

(c) Sketch the potential using the p^+-emitter region as $V = 0$.

(d) Calculate the total potential difference between the emitter and collector. Is the emitter at a higher or lower potential than the collector?

1.3 A silicon n^+pn BJT is at thermal equilibrium.

$$N_{DE} = 10^{17}, \qquad N_{AB} = 10^{16}, \qquad N_{DC} = 10^{15}/cm^3$$

Let $n_i = 10^{+10}/cm^3$ and $kT = 0.026$ eV.

(a) Sketch the energy band diagram.

(b) Sketch the charge density and electric field.

(c) Sketch the potential using the n^+-emitter region as $V = 0$.

(d) Calculate the total potential difference between the emitter and collector.

(e) Is the emitter at a higher or lower potential than the collector?

1.4 If in Problem 1.2 $V_{EB} = 0.5$ volt and $V_{CB} = -2$ volts,

(a) Sketch the energy band diagram with respect to the thermal equilibrium diagram.

(b) What is the potential barrier for the diffusion of holes from the collector to the base?

(c) What is the potential barrier for the diffusion of holes from the emitter to the base?

1.5 If the p^+np BJT is operating only slightly ($I_C > 0$) in the saturation region, make a figure similar to Fig. 1.6 showing the major carrier fluxes and currents.

1.6 If the n^+pn BJT is operating in the cutoff region, make a figure similar to Fig. 1.6 showing the major carrier fluxes and currents.

1.7 For a *pnp* device with $I_{Ep} = 1$ mA, $I_{En} = 0.01$ mA, $I_{Cp} = 0.98$ mA, and $I_{Cn} = 0.1$ μA, calculate

(a) Base transport factor

(b) Emitter injection efficiency

(c) α_{dc} and β_{dc}; the value of I_B

(d) I_{BCO} and I_{ECO}

(e) If $I_{Cp} = 0.99$ mA, calculate β_{dc} and I_B.

(f) If $I_{Cp} = 0.99$ mA and $I_{En} = 0.005$ mA, calculate β_{dc} and I_B.

(g) How will β_{dc} change if I_{En} is increased?

1.8 If in Problem 1.2 the metallurgical base width $W_{BB} = 2$ μm,

(a) What is the quasi-neutral base width, W?

(b) How much collector-to-base voltage can be applied before the base has been eliminated? Assume that $V_{BE} = 0$ volts.

1.9 For an *npn* device with $I_{En} = 100$ μA, $I_{Ep} = 1$ μA, $I_{Cn} = 99$ μA, and $I_{Cp} = 0.1$ μA, calculate

(a) Base transport factor

(b) Emitter injection efficiency

(c) α_{dc} and β_{dc}; the value of I_B

(d) I_{CBO} and I_{CEO}

(e) If $I_{Cn} = 99.5$ μA, calculate β_{dc} and I_B.

(f) If $I_{Cn} = 99$ μA and $I_{Ep} = 2$ μA, calculate β_{dc} and I_B.

(g) How will β_{dc} change if I_{En} is increased?

1.10 If for a *pnp* BJT the base width, W, is much much smaller than the minority carrier diffusion length in the base, let $V_{CB} = 0$ and derive an equation for the emitter injection efficiency. Assume that the emitter is infinite in length and that all the junctions are step junctions with uniformly doped bulk regions. [*Hint*: remember the "short base diode." If the base width were made even smaller, how would this affect β_{dc}?]

1.11 Explain why the current I_{BCO} has no appreciable component of hole current from the base to the collector in the p^+np device.

2 / The Ideal Junction Transistor

The modern transistor has very narrow base region width and therefore little recombination in the base quasi-neutral bulk region. We define the *ideal bipolar junction transistor* as a device with no recombination–generation in the base bulk (quasi-neutral) region, the E–B depletion region, and the C–B depletion region. Also assumed are low-level injection, no electric fields in the bulk regions, one-dimensional current flow, and no external sources of generation such as light. Equations to be solved for the case of recombination in the base bulk region are presented here but will be solved in Chapter 3.

The purpose of Chapter 2 is to use the bipolar transistor model of Chapter 1 to develop quantitative expressions relating the terminal currents to the applied voltages between the E–B and B–C terminals. The derivation is preceded by a coherent "game plan" so that the reader has less chance of straying from the fold. Our game plan develops the necessary equations and discussions without going through the detailed mathematical calculations. Once the game plan is well in hand the derivation of the ideal BJT is developed by starting with the solution for the minority carrier concentration in the base region. After the minority carrier concentrations in the bulk regions are obtained, they are used to obtain the terminal currents. Throughout the development the reader should be aware of the similarity of this derivation to that of the *p-n* diode current. The chapter ends with the definition of the Ebers–Moll equations for the BJT, which are often the basis for the BJT models used in computer aided design (CAD) circuit analysis programs.

2.1 QUANTITATIVE ANALYSIS: GAME PLAN

To obtain the theoretical current-voltage expressions with a minimum amount of confusion, let us assume that the transistor has the currents and voltages as shown in Fig. 2.1(a) for the *pnp* and Fig. 2.1(b) for the *npn*. Other assumptions which we will use either explicitly or implicitly are: (1) the structure is one-dimensional (all internal varibles depend only on x); (2) the bulk base width, W, is less than a minority carrier diffusion length; (3) recombination–generation currents arising in the depletion regions are negligible (the same assumption we used in establishing the ideal diode equation);

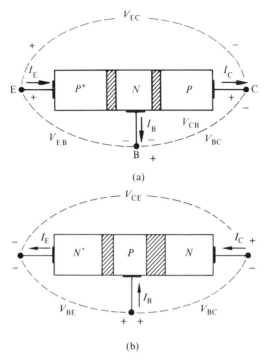

Fig. 2.1   Current and voltage definitions for (a) p^+np and (b) n^+pn.

(4) uniform doping density in the emitter, collector, and base; and (5) low-level injection in all bulk regions. Note from Fig. 2.2 that the zero axis of the coordinates, $x = 0$, has been chosen on the bulk region base side of the E–B depletion region. The base width, W, as previously defined in Fig. 1.6(b), is the quasi-neutral part of the total base width. The x' and x'' axes for the collector and emitter regions, respectively, are also chosen to start at the depletion region edges.

The "game plan" applies to any bipolar transistor derivation and will also be used in Chapter 3 for the nonideal device. With modifications of the symbols it can be used for the *npn*. We emphasize the p^+np transistor here because of its E–B being very similar to the p^+-n diode of Volume II. It will make the subject easier for the reader to follow even though n^+pn devices are more common in circuits. Remember that the "complement" of the *pnp* is the *npn*.

2.1.1   Notation

The two *p*-regions of the *pnp* transistor result in a notational problem. We are not able to use the same symbol for the electron minority carrier diffusion length, L_N, in the emitter and the collector because with different doping L_N may have two different

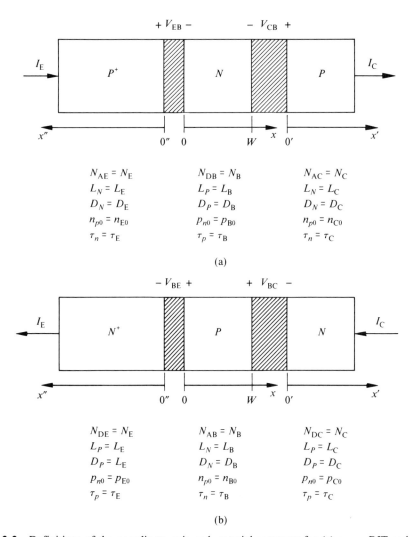

Fig. 2.2 Definitions of the coordinate axis and material constants for (a) a *pnp* BJT and (b) an *npn* BJT.

values. A generally accepted set of symbols is illustrated in Fig. 2.2(a). *The symbol always refers to the minority carrier in the region.* For example, L_N in the emitter becomes L_E, meaning the diffusion length for the minority carrier in the emitter, in this case for the *pnp*, the electron diffusion length. For the collector the symbol is L_C. For the base, L_B is the minority carrier diffusion length in the base; for the *pnp* this means $L_P = L_B$. Similar problems occur for the diffusion constants D_N and D_P as well as for the thermal equilibrium carrier concentrations n_{p0} and p_{n0}. Figure 2.2(a) introduces the symbol n_{E0} for the thermal equilibrium, minority carrier concentration in the emitter and n_{C0} for the collector, while D_E, D_C, and τ_C are used for the minority electron diffusion constants and lifetimes. The base has L_B, D_B, τ_B, and p_{B0} for its minority carrier symbols. Note that a similar set of symbols can be defined for the *npn*, as shown in Fig. 2.2(b).

2.1.2 Current Definitions

The ideal diode equation derivation used the minority carrier currents at the edges of the depletion region to determine the total current through the junction. Paralleling the ideal diode equation derivation, with no generation–recombination in the E–B or C–B depletion regions, the electron and hole currents entering a depletion region are constant throughout the region. What goes in must come out. Figure 2.3 illustrates the current components for the E–B and C–B depletion regions. The emitter current crossing the E–B depletion region is evaluated as *two minority carrier diffusion currents* [Eqs. (2.1) and (2.2)], where the holes injected from the emitter into the base constitute the first current term and the back-injected electrons from the base to the emitter constitute the second.

$$I_E = I_{Ep}(0) + I_{En}(0'') \tag{2.1}$$

$$I_E = -qAD_B \frac{d\,\Delta p_B}{dx}\bigg|_{x=0} - qAD_E \frac{d\,\Delta n_E}{dx''}\bigg|_{x''=0} \tag{2.2}$$

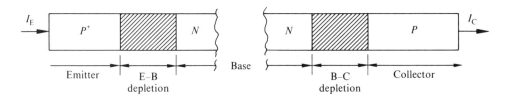

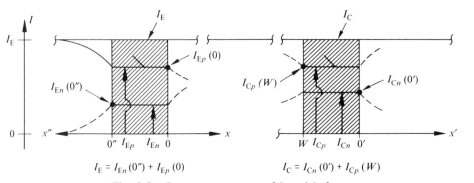

Fig. 2.3 Current components of I_E and I_C for a *pnp*.

Note that each term is evaluated at the edge of the depletion region. The minus sign in the second term in Eq. (2.2) is a result of the change of axis for electrons in the emitter. The emitter electron current is positive when flowing in the direction of the x''-axis since I_E is positive in the opposite direction (the positive x-axis). Note that the gradient term $d \, \Delta n_E / dx''$ is negative.

The collector current components at the C–B junction are determined by the holes entering the depletion region from the base side and by the electrons from the p-side. Again, each *minority carrier diffuson current* is evaluated at the depletion region edges; that is,

$$I_C = I_{Cp}(W) + I_{Cn}(0') \tag{2.3}$$

$$I_C = -qAD_B \frac{d \, \Delta p_B}{dx}\bigg|_{x=W} + qAD_C \frac{d \, \Delta n_C}{dx'}\bigg|_{x'=0} \tag{2.4}$$

The similarity between these definitions for junction transistor currents and the ideal diode current derivation should be evident. If it is not, Chapter 3 of Volume II should be reviewed immediately!

2.1.3 Bulk Region Solution

Equations (2.1) through (2.4) emphasize the need to know the minority carrier concentrations in each bulk region. The minority carrier concentrations $n_E(x'')$, $p_B(x)$, and $n_C(x')$ need to be derived. The solution to the minority carrier continuity equation, assuming

1. there is low-level injection,
2. the electric field for the minority carrier is zero,
3. there is no generation due to light, etc., and
4. a steady-state, dc solution,

yields the *minority carrier diffusion equations* of Volume I. Equation (2.5) is the equation to be solved for holes in the base region of a *pnp* device where $\tau_p = \tau_B$ by our new notation:

$$D_B \frac{d^2 \, \Delta p_B(x)}{dx^2} = \frac{\Delta p_B(x)}{\tau_B} \tag{2.5}$$

Since the doping of the base is uniform, p_{B0} is constant and any x variation must be in Δp_B. Equation (2.5) is written as Eq. (2.6) by dividing both sides by D_B and defining the minority carrier diffusion length L_P as $L_B = \sqrt{D_B \tau_B}$:

$$\frac{d^2 \, \Delta p_B(x)}{dx^2} = \frac{\Delta p_B(x)}{L_B^2} \tag{2.6}$$

which has a solution of the form

$$\Delta p_{B}(x) = C_1 e^{x/L_B} + C_2 e^{-x''/L_B} \qquad (2.7a)$$

where two boundary conditions are needed to determine C_1 and C_2.

Similar arguments result in minority carrier diffusion equations for $\Delta n_E(x'')$ and $\Delta n_C(x')$, with $L_E = \sqrt{D_E \tau_E}$ and $L_C = \sqrt{D_C \tau_C}$, respectively. Each diffusion equation is a second-order linear differential equation and requires two boundary conditions for a complete solution. For the emitter bulk region,

$$\Delta n_E(x'') = C_1 e^{x''/L_E} + C_2 e^{-x''/L_E} \qquad (2.7b)$$

and for the collector bulk region,

$$\Delta n_C(x') = C_1 e^{x'/L_C} + C_2 e^{-x'/L_C} \qquad (2.7c)$$

2.1.4 Boundary Conditions

The junction voltages V_{EB} and V_{CB} establish the minority carrier concentrations at the edges of the depletion regions. The derivation is identical to that for the p-n junction diode of Volume II, and we present only the results in Eqs. (2.8) through (2.11) for the E–B and C–B junctions.

$$\Delta n_E(0'') = n_{E0}(e^{qV_{EB}/kT} - 1) \qquad (2.8)$$

$$\Delta p_B(0) = p_{B0}(e^{qV_{EB}/kT} - 1) \qquad (2.9)$$

$$\Delta p_B(W) = p_{B0}(e^{qV_{CB}/kT} - 1) \qquad (2.10)$$

$$\Delta n_C(0') = n_{C0}(e^{qV_{CB}/kT} - 1) \qquad (2.11)$$

Figure 2.4 illustrates these boundary conditions in terms of the total carrier concentrations with the E–B forward biased and the C–B reverse biased; that is, active region operation.

The additional boundary conditions required for a solution must come from the regions away from the depletion regions. For the case where the emitter bulk region is much, much longer than the minority carrier diffusion length L_E, $\Delta n_E(x'' \rightarrow \infty) = 0$ since eventually all the injected electrons must recombine with majority carrier holes. Similarly for the collector, $\Delta n_C(x' \rightarrow \infty) = 0$. Remember that "$L$" is the average distance the minority carrier will diffuse before recombining with a majority carrier. Therefore, if the emitter and collector bulk regions are much longer than the minority carrier diffusion length, the minority carrier concentration reaches its thermal equilibrium value at large distances from the junction, as illustrated in Fig. 2.4.

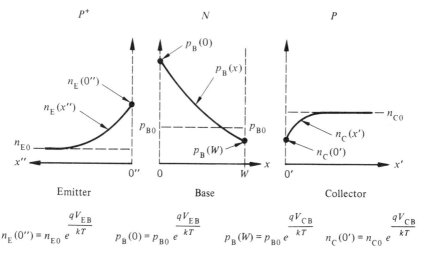

Fig. 2.4 Minority carrier concentrations and boundary conditions in the bulk portions of a device operating in the active region.

2.1.5 Emitter Current

To obtain an equation for the emitter current, I_E, we need to determine I_{Ep} and I_{En} by performing the following steps:

1. Solve the minority diffusion equation in the base, Eq. (2.6), for $\Delta p_B(x)$.
2. Apply the boundary conditions $\Delta p_B(0)$ and $\Delta p_B(W)$, Eqs. (2.9) and (2.10), to Eq. (2.7a) and evaluate the two constants C_1 and C_2.
3. Solve the minority carrier diffusion equation in the emitter bulk region, Eq. (2.12), for $n_E(x'')$, Eq. (2.7b):

$$\frac{d^2 \Delta n_E(x'')}{dx''^2} = \frac{\Delta n_E(x'')}{L_E^2} \tag{2.12}$$

where $L_E = \sqrt{D_E \tau_E}$.

4. Apply the boundary conditions, namely $\Delta n_E(x'' = \infty) = 0$ and $n_E(0'')$; that is, Eq. (2.8).
5. Evaluate Eq. (2.2) for the diffusion current components at each edge of the depletion region as indicated in Fig. 2.3, which will yield I_E as a function of V_{EB} and V_{CB}.

2.1.6 Collector Current

To obtain an equation for the collector current, I_C, the following steps are performed to solve for I_{Cp} and I_{Cn}:

1. Solve a diffusion equation in the bulk collector region, Eq. (2.13), for $\Delta n_C(x')$, Eq. (2.7c):

$$\frac{d^2 \Delta n_C(x')}{dx'^2} = \frac{\Delta n_C(x')}{L_C^2} \tag{2.13}$$

where $L_C = \sqrt{D_C \tau_C}$.

2. Apply the boundary conditions, namely $\Delta n_C(x' = \infty) = 0$ and $\Delta n_C(0')$, Eq. (2.11).

3. Evaluate Eq. (2.4) for the current components at each edge of the B–C depletion region as shown in Fig. 2.3, which will yield I_C as a function of V_{EB} and V_{CB}.

2.1.7 Base Current

To solve for I_B, apply Kirchhoff's Current Law, which states that

$$\boxed{I_B = I_E - I_C} \tag{2.14}$$

We could also obtain I_B from the equations by noting that $I_{B1} + I_{B2} - I_{B3} = I_{En} + (I_{Ep} - I_{Cp}) - I_{Cn}$.

2.2 THE IDEAL BIPOLAR TRANSISTOR

As a special case of the bipolar transistor, we define the *ideal bipolar transistor* as having a base bulk region so narrow that no recombination (or generation) occurs in the base. In modern devices the quasi-neutral base width (W) is often less than 1 μm while the minority carrier diffusion length is about 20 to 35 μm. With such narrow base widths the minority carriers do not stay in the region long enough to have a significant chance of recombination. Therefore, we approximate the real device with the ideal case.

A second reason for starting with an ideal device is the greatly simplified equations that result for I_E and I_C where the current components can easily be identified and related to the diode currents. In addition, several concepts can be easily demonstrated with the ideal model without getting lost in the mathematical detail.*

If no recombination of holes is to occur in the base region of a *pnp*, the lifetime $\tau_p = \tau_B$ must approach infinity; that is, L_B must approach infinity, or alternatively, $W \ll L_B$. When applied to Eq. (2.6), the right side of the equation becomes zero:

$$\frac{d^2 \Delta p_B(x)}{dx^2} \cong 0 \tag{2.15}$$

---

*It avoids writing $p_B(x)$ in terms of hyperbolic functions; see Chapter 3.

Integrating twice, remembering the constants of integration, yields

$$\Delta p_B(x) = C_1 x + C_2 \tag{2.16}$$

Note that Eq. (2.16) is the equation of a straight line. Applying the two boundary conditions at $x = 0$ and $x = W$ to Eq. (2.16) yields C_2 and then C_1, respectively:

$$\Delta p_B(0) = C_1[0] + C_2 = C_2 \tag{2.17}$$

$$\Delta p_B(W) = C_1 W + C_2 = C_1 W + \Delta p_B(0) \tag{2.18}$$

Therefore,

$$C_1 = \frac{\Delta p_B(W) - \Delta p_B(0)}{W}$$

We can write the $\Delta p_B(x)$ equation as

$$\Delta p_B(x) = -\left[\frac{\Delta p_B(0) - \Delta p_B(W)}{W}\right]x + \Delta p_B(0) \tag{2.19}$$

which shows the straight line equation more explicitly.

Figure 2.5 illustrates the minority carrier concentrations for the ideal transistor in the active, saturation, and cutoff regions of operation. Note that each case depends

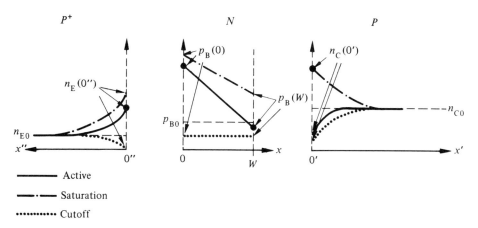

——— Active

—·— Saturation

··········· Cutoff

Fig. 2.5 Ideal *pnp* minority carrier concentrations, with no recombination in the base bulk region.

on the polarities of V_{EB} and V_{CB} since $\Delta p_B(0) = p_{B0}(e^{qV_{EB}/kT} - 1)$ and $\Delta p_B(W) = p_{B0}(e^{qV_{CB}/kT} - 1)$.

A quick inspection of Fig. 2.5 and Eq. (2.19) indicates that the slope of $\Delta p_B(x)$ is

$$-\left[\frac{\Delta p_B(0) - \Delta p_B(W)}{W}\right] = \text{slope} \qquad (2.20)$$

which is used in calculating the hole current entering and leaving the base region as indicated by Eqs. (2.2) and (2.4). Since the slope is constant for the ideal device, the hole current is constant in the base region, which must be the case for no hole recombination. The hole particle flux into the base region from the emitter equals the hole particle flux out of the base region into the collector. Comparing Fig. 2.5 with Fig. 2.6, the reader should note the difference in $\Delta p_B(x)$. For a device with significant recombination, the plot is not a straight line; this indicates that holes are lost (recombined) in W. Also note that the slopes of $\Delta p_B(x)$ at $x = 0$ and at $x = W$ are different, indicating that holes are recombined (lost) in traversing the base bulk region and that $I_{Cp} < I_{Ep}$.

2.3 THE IDEAL *pnp* TRANSISTOR CURRENTS

This section carries out the mathematical derivations for the ideal *pnp* bipolar transistor to obtain equations for the terminal currents as functions of the junction voltages and the material parameters such as doping and diffusion lengths. We are applying the "game plan" as outlined in Section 2.1 to the ideal device of Section 2.2.

To evaluate Eq. (2.2) for the emitter current, we need first to solve Eq. (2.12) for $\Delta n_E(x'')$ in the bulk emitter region. The solution of the minority carrier diffusion equation is repeated here as Eq. (2.21):

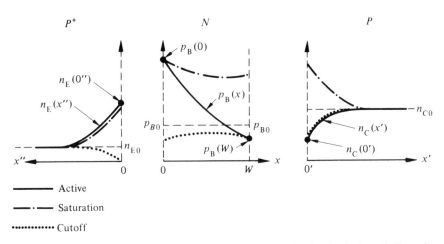

Fig. 2.6 *pnp* minority carrier concentrations with recombination in the base bulk region.

$$\Delta n_E(x'') = C_1 e^{-x''/L_E} + C_2 e^{x''/L_E} \tag{2.21}$$

Two boundary conditions are needed to solve for C_1 and C_2. Since an electron cannot survive forever in a *p*-region, $\Delta n_E(x'' = \infty) = 0$ and therefore C_2 must be zero. At $x'' = 0''$,

$$\Delta n_E(0'') = C_1 e^{0''} = C_1 = n_{E0}(e^{qV_{EB}/kT} - 1) \tag{2.22}$$

from Eq. (2.8) for the boundary condition. Hence the solution is

$$\Delta n_E(x'') = n_{E0}(e^{qV_{EB}/kT} - 1)e^{-x''/L_E} \tag{2.23}$$

Equation (2.23) is plotted in Figs. 2.4 and 2.5. The back injected electron current from the base to the emitter, $I_{B1} = I_{En}(0'')$ (as sketched in Fig. 2.3), can now be evaluated from Eq. (2.2):

$$I_{En}(0'') = -qAD_E \frac{d\,\Delta n_E(x'')}{dx}\bigg|_{x''=0''} = \frac{qAD_E}{L_E} n_{E0}(e^{qV_{EB}/kt} - 1) \tag{2.24}$$

The hole current component $I_{Ep}(0)$ is easily evaluated for the ideal device because the slope has already been determined as Eq. (2.20):

$$I_{Ep}(0) = -qAD_B \frac{d\,\Delta p_B}{dx}\bigg|_{x=0} = \frac{qAD_B}{W}[\Delta p_B(0) - \Delta p_B(W)] \tag{2.25}$$

Substituting the boundary conditions for $\Delta p_B(0)$ and $\Delta p_B(W)$, Eqs. (2.9) and (2.10), yields

$$I_{Ep}(0) = \frac{qAD_B}{W} p_{B0}[(e^{qV_{EB}/kT} - 1) - (e^{qV_{CB}/kT} - 1)] \tag{2.26}$$

Equation (2.2) shows the emitter current to be the sum of Eqs. (2.24) and (2.26), $I_E = I_{Ep} + I_{En}$:

$$\boxed{I_E = qA\left[\frac{D_E n_{E0}}{L_E} + \frac{D_B p_{B0}}{W}\right](e^{qV_{EB}/kT} - 1) - \left[\frac{qAD_B}{W} p_{B0}\right](e^{qV_{CB}/kT} - 1)}$$

$$\tag{2.27}$$

Note how the first term of Eq. (2.2) resembles that of the "short base" diode of several problems in Volume II; if $V_{CB} = 0$, it is the same mathematical form.

The collector current (I_C) can be derived by a similar set of calculations. However, we should be able to evaluate the first term of Eq. (2.4) directly. Since the hole current I_{Ep} is a constant in the base (no hole recombination in the base), its value entering the base must be the same as that leaving the base; that is,

$$I_{Ep}(0) = I_{Ep}(W) = I_{Cp} \tag{2.28}$$

The current component $I_{Cn}(0')$ can be obtained directly from Eq. (2.24) by exchanging "E" for "C" and V_{EB} for V_{CB}.* A change in sign is necessary because in this case I_C is in the same direction as the x' axis, which was not the case for the emitter:

$$I_{Cn} = \frac{-qAD_C}{L_C} n_{C0}(e^{qV_{CB}/kT} - 1) \tag{2.29}$$

The collector current is the sum of Eqs. (2.26) and (2.29). Hence,

$$I_C = \left[\frac{qAD_B}{W} p_{B0}\right](e^{qV_{EB}/kT} - 1) - qA\left[\frac{D_C n_{C0}}{L_C} + \frac{D_B p_{B0}}{W}\right](e^{qV_{CB}/kT} - 1) \tag{2.30}$$

Alternatively, Eq. (2.13) could be solved subject to the boundary conditions $\Delta n_C(x' \to \infty)$ and $\Delta n_C(0')$. Then apply the second term of Eq. (2.4) to get Eq. (2.29). Refer to Exercise 2.1 in Appendix A. The second term of Eq. (2.30) is very similar to the "short base" n-p junction diode. In fact, if $V_{EB} = 0$, it is identical.

The base current for the ideal transistor is obtained as $I_B = I_E - I_C$. Subtracting Eq. (2.30) from Eq. (2.27) yields

$$I_B = \frac{qAD_E}{L_E} n_{E0}(e^{qV_{EB}/kT} - 1) + \frac{qAD_C}{L_C} n_{C0}(e^{qV_{CB}/kT} - 1) \tag{2.31}$$

Equations (2.27), (2.30), and (2.31) are applicable to all regions of operation for the ideal BJT; that is, active, saturation, cutoff, and inverted.

The terminal currents for the BJT can be expressed in a slightly different but equivalent manner. Since $n_{E0} = n_i^2/N_{AE} = n_i^2/N_E$, $n_{B0} = n_i^2/N_{DB} = n_i^2/N_B$, and $n_{C0} = n_i^2/N_{AC} = n_i^2/N_C$, Eqs. (2.27), (2.30), and (2.31) can also be written as follows:

$$I_E = qAn_i^2\left[\frac{D_E}{L_E N_E} + \frac{D_B}{WN_B}\right](e^{qV_{EB}/kT} - 1) - qAn_i^2\left[\frac{D_B}{WN_B}\right](e^{qV_{CB}/kT} - 1) \tag{2.32a}$$

*Alternatively, Eq. (2.13) could be solved subject to the boundary conditions $\Delta n_C(x' = \infty)$ and $\Delta n_C(0')$. Then apply the second term of Eq. (2.4) to get Eq. (2.29).

$$I_C = qAn_i^2 \left[\frac{D_B}{WN_B} \right] (e^{qV_{EB}/kT} - 1) - qAn_i^2 \left[\frac{D_C}{L_C N_C} + \frac{D_B}{WN_B} \right] (e^{qV_{CB}/kT} - 1)$$

(2.32b)

$$I_B = qAn_i^2 \left[\frac{D_E}{L_E N_E} \right] (e^{qV_{EB}/kT} - 1) + qAn_i^2 \left[\frac{D_C}{L_C N_C} \right] (e^{qV_{CB}/kT} - 1)$$

(2.32c)

These forms of the currents I_E, I_C, and I_B show the explicit dependence on n_i^2 which increases exponentially with an increase in temperature.

The base current, I_B, Eq. (2.31) or (2.32c), is made up of two terms that have the same mathematical form, indicating a symmetry if the emitter and collector were to be interchanged.

SEE EXERCISE 2.1–APPENDIX A

2.3.1 Active Region

The purpose of this section is to show the major current components of the BJT when operating in the middle of the active region. Later these results will be used to determine β_{dc} and α_{dc} in terms of the material parameters.

Equations (2.27), (2.30), and (2.31) confirm the qualitative analysis of the active region *pnp* presented previously in Fig. 1.6(b). For active region operation, V_{CB} is negative and the exponential terms in these equations are, for only a few tenths of a volt negative, much less than one, because q/kT is about 38.46 at room temperature; that is,

$$\exp\left(\frac{qV_{CB}}{kT}\right) \ll 1, \quad \text{and, similarly} \quad \exp\left(\frac{qV_{EB}}{kT}\right) \gg 1$$

because V_{EB} is positive and by as little as a few tenths of a volt the exponential is much larger than one. Therefore we can ignore the exponential in the first case and the "1" in the second. For the emitter current, note that in Eq. (2.27) or (2.32a) they could be approximated as

$$I_E \cong qAn_i^2 \left[\frac{D_E}{L_E N_E} + \frac{D_B}{WN_B} \right] (e^{qV_{EB}/kT})$$

where the "-1" terms have cancelled each other. Figure 2.7 illustrates the active region current components. We can identify the current components as

$$I_{Ep} = qAn_i^2 \frac{D_B}{WN_B} e^{qV_{EB}/kT}$$

which represents the holes injected from the emitter into the base. For the ideal BJT,

$$I_{Cp} \equiv I_{Ep}$$

since no holes are lost due to recombination in the base bulk region, W. The electrons back injected from the base into the emitter are

$$I_{En} \cong qAn_i^2 \frac{D_E}{L_E N_E} e^{qV_{EB}/kT}$$

where $I_{En} = I_{B1}$ of the base current. Hence,

$$I_E \cong I_{Ep} + I_{En} \cong qAn_i^2 \left[\frac{D_E}{L_E N_E} + \frac{D_B}{WN_B} \right] e^{qV_{EB}/kT} \qquad (2.33a)$$

Note the exponential dependence of the emitter current on V_{EB}.

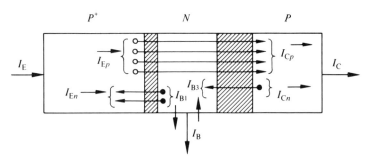

Fig. 2.7   Current components of an ideal p^+np operating in the active region.

The collector current is made up of $I_{Ep} = I_{Cp}$ and I_{Cn}, which also becomes I_{B3} (the thermally generated minority carrier electrons in the collector). Note that in Eq. (2.30) or (2.32b) the $\exp(qV_{CB}/kT) \ll 1$; then,

$$I_{Cn} \cong qAn_i^2 \left[\frac{D_C}{L_C N_C} \right] = I_{B3}$$

and the collector current in the active region is approximated as

$$I_C \cong qAn_i^2 \left[\frac{D_B}{WN_B} \right] e^{qV_{EB}/kT} + qAn_i^2 \left[\frac{D_C}{L_C N_C} \right]$$

which for a reasonable level of V_{EB} becomes

$$I_C \cong qAn_i^2 \frac{D_B}{WN_B} e^{qV_{EB}/kT} \qquad (2.33b)$$

because the exponential term is the much larger of the two.

The base current, by inspection of Eq. (2.31) or (2.32c), becomes

$$I_B \cong qAn_i^2 \frac{D_E}{L_E N_E} e^{qV_{EB}/kT} - \frac{qAn_i^2 D_C}{L_C N_C}$$

The first term is I_{En} and the second is $I_{Cn} = I_{B3}$. Since the exponential term is large,

$$I_B \cong qAn_i^2 \frac{D_E}{L_E N_E} e^{qV_{EB}/kT} \qquad (2.33c)$$

Active region operation can, to a first order, be said to be controlled by V_{EB} and is relatively independent of V_{CB} as long as it is negative by a few tenths of a volt.

2.3.2 Saturation Region

The ideal BJT operating in the saturation region has both the E–B and C–B junctions forward biased. Figure 2.8 illustrates the current components and carrier fluxes in saturation. The very popular integrated circuit family transistor-transistor logic (TTL) has devices that operate in this region. At the edge between saturation and active region

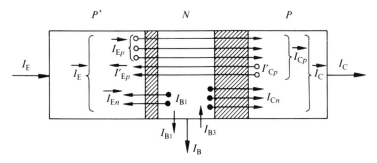

Fig. 2.8   Major current components of an ideal p^+np in the saturation region.

operation, $V_{CB} = 0$ and $V_{EB} > 0$. Further into saturation, $V_{CB} > 0$ and $V_{EB} > 0$; therefore both exponential terms are large compared to the "1" and hence

$$e^{qV_{EB}/kT} \gg 1 \quad \text{and} \quad e^{qV_{CB}/kT} \gg 1.$$

Inspection of Eq. (2.32a) for I_E shows that

$$I_E \cong \underbrace{qAn_i^2 \frac{D_E}{L_E N_E} e^{qV_{EB}/kT}}_{I_{En}} + \underbrace{qAn_i^2 \frac{D_B}{WN_B} e^{qV_{EB}/kT}}_{I_{Ep}} - \underbrace{qAn_i^2 \frac{D_B}{WN_B} e^{qV_{CB}/kT}}_{I'_{Cp}} \qquad (2.34a)$$

In this case the C–B is forward biased and is injecting holes from the collector (I'_{Cp}) to the emitter; hence the minus sign, since the flow is opposite to the holes injected from the emitter. If $V_{EB} = V_{CB}$, the first term of Eq. (2.34a) is left, showing that I_E is still positive. Figure 2.8 illustrates the three major current components of Eq. (2.34a). Note that if $V_{CB} > V_{EB}$ by a large enough amount the emitter current could become negative! For the collector current in saturation, Eq. (3.32b) becomes

$$I_C \cong \underbrace{qAn_i^2 \left[\frac{D_B}{WN_B}\right] e^{qV_{EB}/kT}}_{I_{Ep}} - \underbrace{qAn_i^2 \frac{D_C}{L_C N_C} e^{qV_{CB}/kT}}_{I_{B3} = -I_{Cn}} - \underbrace{qAn_i^2 \frac{D_B}{WN_B} e^{qV_{CB}/kT}}_{I'_{Cp}} \qquad (2.34b)$$

Note that $I_{Cp} = I_{Ep} + I'_{Cp}$ and that $I_C = I_{Cp} + I_{Cn}$. Here we see that the electrons are injected from the base to the collector since that junction is forward biased. The current I_{B3} has reversed direction and become much larger and may be comparable in size to I_{B1}.

Base current is now much larger than in the active region since it must supply electrons to both the E–B and C–B junctions. From Eq. (2.32c),

$$I_B \cong qAn_i^2 \frac{D_E}{L_E N_E} e^{qV_{EB}/kT} + qAn_i^2 \frac{D_C}{L_C N_C} e^{qV_{CB}/kT} \qquad (2.34c)$$

$$\underbrace{\phantom{qAn_i^2 \frac{D_E}{L_E N_E} e^{qV_{EB}/kT}}}_{I_{B1} = I_{En}} \qquad \underbrace{\phantom{qAn_i^2 \frac{D_C}{L_C N_C} e^{qV_{CB}/kT}}}_{-I_{B3} = -I_{Cn}}$$

We see that I_{B3} is negative and, with a large enough V_{BC}, comparable in magnitude to I_{B1}. The result is that I_B has increased considerably in the saturation region even if I_{B1} is about the same as before.

2.3.3 Cutoff Region

Cutoff region operation requires that both the E–B and C–B junctions be reverse biased. Figure 2.9 illustrates the major current components and carrier fluxes. Remember that the exponentials are much less than one for negative V_{EB} and V_{CB} of a few tenths of a volt:

$$e^{qV_{EB}/kT} \ll 1 \qquad \text{and} \qquad e^{qV_{CB}/kT} \ll 1$$

From Eqs. (2.32a), (2.32b), and (2.32c), the terminal currents are approximated as

$$I_E \cong -qAn_i^2 \frac{D_E}{L_E N_E} \qquad (2.35a)$$

$$I_C \cong qAn_i^2 \frac{D_C}{L_C N_C} \qquad (2.35b)$$

$$I_B \cong -qAn_i^2 \frac{D_E}{L_E N_E} - qAn_i^2 \frac{D_C}{L_C N_C} \qquad (2.35c)$$

We see that for I_E the thermally generated minority carrier electrons in the emitter drift down the potential hill into the base and therefore $I_E < 0$. For the collector the ther-

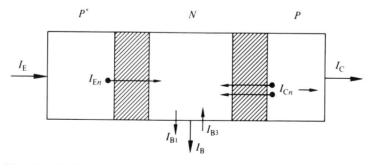

Fig. 2.9   Major current components of an ideal p^+np in the cutoff region.

mally generated electrons also fall down the potential hill of the C–B and hence $I_{Cn} = I_C > 0$. Since $N_{AC} \ll N_{AE}$, more minority carriers are generated in the collector than in the emitter. Also note that both components of the base current are negative.

2.3.4   Inverted Operation

The inverted operation occurs when the V_{CB} becomes large enough positive to reverse the net collector current; i.e., it becomes negative. Inverted saturation and inverted active operation can occur depending on the value of V_{EB}. If positive, forward biased, the BJT operates in the inverted saturation region. If V_{EB} is negative, then inverted active operation occurs. Figure 2.10 illustrates the major carrier flux and currents for inverted-active operation.

The current components I_{Cp}, I_{Cn}, I_{En}, and I_{Ep} are all negative, as are I_{B1} and I_{B3}. Because I_{B3} is negative and large in magnitude, the net base current is positive since $I_B = I_{B1} - I_{B3}$, as shown in Fig. 1.3(b). Note that the collector-to-base injection efficiency is low because $N_{AC} < N_{DB}$. Therefore, more electrons are injected from base to emitter than holes from collector to base.

Considering Eq. (2.32b) with $V_{EB} < 0$ and $V_{CB} > 0$, the collector current becomes, in the inverted active mode,

$$I_C \cong -qAn_i^2 \left[\frac{D_C}{L_C N_C} + \frac{D_B}{W N_B} \right] e^{qV_{CB}/kT} \qquad (2.36a)$$

which shows that $I_C < 0$ and is controlled by the V_{CB} voltage. The first term inside the brackets of Eq. (2.36a) represents the I_{Cn} electrons injected from the base to the collector, and the second term represents I_{Cp} due to holes injected from the collector to the base. Similarly for the emitter current we see that

$$I_E \cong -qAn_i^2 \left[\frac{D_B}{W N_B} \right] e^{qV_{CB}/kT} \qquad (2.36b)$$

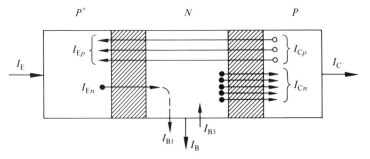

Fig. 2.10   Inverted active operation major carrier flux for the p^+np device.

The base current in the inverted active mode, from Eq. (2.32c), is

$$I_B \cong qAn_i^2 \left[\frac{D_C}{L_C N_C} \right] e^{qV_{CB}/kT} \qquad (2.36c)$$

which shows it to be positive, and the major component is due to the electrons injected from the base to the collector.

There are several applications in digital and analog circuits that use the inverted mode of operation. Note that γ, α_{dc}, and β_{dc} can now be defined as γ_R, α_{dcR}, and β_{dcR}, all of which are lower in inverted active operation.

2.4 IDEAL *V–I* CHARACTERISTICS

A major purpose of this section is to obtain equations for α_{dc} and β_{dc} in terms of the doping densities, other material parameters, and device dimensions. Using this information we can effectively design devices for better circuit performance. A second objective is to plot the BJT input and output *V–I* characteristics.

The current equations (Eqs. 2.32a, b, and c) obtained in the previous sections can be used to plot the input and output characteristics for the common base and common emitter configurations. (See Fig. 1.4 for definitions of the input and output variables.) A common base amplifier fits the form of the equations very well since V_{EB} and V_{CB} are the input and output voltages and I_E and I_C are the input and output currents. However, the common emitter is the most often used form, with V_{EB} and I_B as the input parameters and V_{EC} and I_C the output variables.

2.4.1 Common Base

Figure 2.11(a) illustrates the ideal BJT input characteristics and Fig. 2.11(b) the output characteristics for the common base configuration. Note that for $V_{CB} = 0$ the input

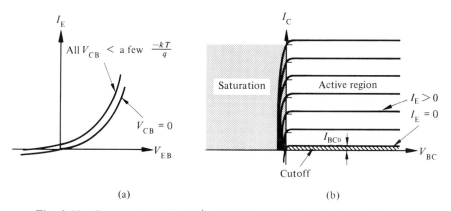

(a)   (b)

Fig. 2.11 Common base ideal p^+np *V–I* characteristics: (a) input; (b) output.

characteristic is similar to that of a forward biased diode (Eq. 2.32a). Shorting the C–B and then using the E–B as a diode is commonplace in integrated circuit designs. For all V_{CB} more negative than several $-kT/q$ volts, the input plot becomes one curve, indicating that the emitter current is primarily comprised of the forward injected holes (which are dependent on V_{EB})—independent of V_{CB}, as shown by Eq. (2.33a) in the active region.

The common base output characteristics of Fig. 2.11(b) indicate that the active region is to the right of the V_{BC} axis (or $V_{CB} < 0$). Note that I_C is independent of V_{BC} for a fixed value of I_E over a large range of V_{BC}.

To the left of the axis, V_{BC} is negative (or $V_{CB} > 0$) and the device is in saturation (for $I_E \geq 0$ and hence $V_{BE} > 0$). In saturation, holes are injected from the collector to the base. These holes are oppositely directed to the holes being injected from the emitter. As a result, the collector hole current components subtract from each other and the *net* collector current decreases as illustrated by the current components in Fig. 2.8. The more forward biased the C–B junction becomes, the greater the collector injection of holes and the smaller the net value of I_C. Note that for larger I_E, larger values of V_{CB} are necessary for significant hole current subtraction and therefore lower I_C.

The cutoff region occurs when both the E–B and C–B junctions are reverse biased. In this case the emitter current is less than zero.

For the ideal device operating in the active region, alpha (α_{dc}) is a measure of how few electrons are back injected from the base to the emitter compared with the number of holes injected from the emitter to the base, as shown in Fig. 2.7. As a performance criterion for the emitter current due to holes, as compared with the total emitter current, the *emitter injection efficiency* (γ) was defined by Eq. (1.5) for a *pnp* transistor in the active region as

$$\gamma_{pnp} = \frac{I_{Ep}}{I_E} = \frac{I_{Ep}}{I_{Ep} + I_{En}}$$

From Eqs. (2.26) and (2.27), the ratio I_{Ep}/I_E, in the active region for the ideal device, can be expressed as in Eq. (2.37), where the exponential terms are cancelled out:

$$\gamma_{pnp} = \frac{\dfrac{D_B p_{B0}}{W}}{\dfrac{D_B p_{B0}}{W} + \dfrac{D_E n_{E0}}{L_E}} \tag{2.37}$$

Note that in Eq. (2.37), as n_{E0} is made much less than p_{B0}, by doping the emitter, $N_{AE} \gg N_{DB}$, then $\gamma \to 1$. If we could, we would make $\gamma = 1$, because this increases the gain of the transistor to phenomenal values (β_{dc} would be infinite).

The *base* transport factor (α_T) is defined by Eq. (1.4); for the ideal device it is unity, because no holes are recombined (lost) as they travel through the base region. Therefore $\alpha_{dc} = \gamma \alpha_T = \gamma$ in this case.

The common base, active region, short circuit current gain I_C/I_E is termed the *dc alpha* (α_{dc}). The ratio I_C/I_E, as defined from Eqs. (1.6) and (1.7), is obtained by applying Eqs. (2.28), (2.26), and (2.24) *in the active region*. Equation (2.38) explicitly shows that W should be small and N_E large:

$$\alpha_{dc} = \frac{\dfrac{D_B p_{B0}}{W}}{\dfrac{D_B p_{B0}}{W} + \dfrac{D_E n_{E0}}{L_E}} = \frac{1}{1 + \dfrac{D_E n_{E0} W}{D_B p_{B0} L_E}} = \frac{1}{1 + \dfrac{D_E N_B W}{D_B N_E L_E}} \tag{2.38}$$

Again we note the need to have $W \ll L_E$ for a high-gain device.

The common base ideal *pnp* transistor output characteristics of Fig. 2.11(b) illustrate the definition of I_{BC0} as the collector current, when the emitter is open circuited ($I_E = 0$) and the C–B junction is reverse biased. Physically, this means that I_{BC0} is the reverse saturation current of the C–B junction, the I_{Cn} component in Fig. 2.7. For a p^+np device, this current is primarily determined by the n_{C0} of the collector because the base is doped more heavily than the collector; and $W \ll L_B$. Hence there is no significant generation of holes in the base. The minority electrons thermally generated within one diffusion length of the C–B depletion region edge drift down the potential hill to the base and become I_{B3} of the base current of Fig. 2.7.

2.4.2 Common Emitter

Figure 2.12(a) illustrates the input characteristics for the ideal common emitter *pnp* bipolar transistor. Note the similarity between the forward biased diode and the input characteristics, I_B versus V_{EB} with $V_{EC} = 0$. When V_{EC} becomes greater than a few

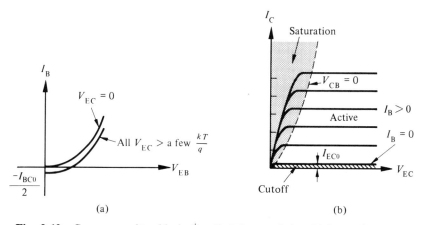

Fig. 2.12   Common emitter ideal p^+np V–I characteristics: (a) input; (b) output.

kT/q, the characteristic becomes independent of V_{EC} because after the B–C junction becomes reverse biased by several kT/q, an additional voltage does not have a major effect on the injection of holes (and electrons) at the E–B junction. Figure 2.8(b) illustrates the output characteristics indicating the active region where, for a fixed I_B, I_C becomes independent of V_{EC} because further reverse bias on the B–C junction does not change the E–B voltage, which is mostly fixed by I_B.

The active region is characterized by a large current gain. This current gain is called the dc beta (β_{dc}) and was defined as I_C/I_B in Eq. (1.11). Because the ideal device in the active region has $V_{CB} < 0$ by several kT/q volts, and because of the non-symmetrical doping of the base and emitter, Eq. (2.33d) for I_C and Eq. (2.33c) for I_B in the active region can be combined with the definition of β_{dc} to give Eq. 2.39, since $\beta_{dc} = I_C/I_B$:

$$I_C \cong qAp_{B0}\frac{D_B}{W}e^{qV_{EB}/kT} \qquad \text{and} \qquad I_B \cong qAn_{E0}\frac{D_E}{L_E}e^{qV_{EB}/kT}$$

$$\beta_{dc} = \frac{qAp_{B0}\dfrac{D_B}{W}e^{qV_{EB}/kT}}{qAn_{E0}\dfrac{D_E}{L_E}e^{qV_{EB}/kT}} \tag{2.39}$$

Therefore, the active region beta (β_{dc}) is defined as

$$\beta_{dc} = \frac{D_B p_{B0}}{D_E n_{E0}}\frac{L_E}{W} = \frac{D_B N_E L_E}{D_E N_B W} \tag{2.40}$$

By doping the emitter heavily (p^+), p_{B0} is made much larger than n_{E0}, and by making W small, the current gain is made large. Again we see why the base region (W) is made very narrow.

If one lets $V_{CB} = 0$, the edge of the active region and saturation region, then the derivation above becomes exact. For $V_{CB} < 0$, the result is a good approximation because the C–B voltage does not significantly affect hole injection at the E–B junction.

The saturation region is defined as the region wherein both the E–B and C–B junctions are forward biased. As V_{CB} becomes more forward biased, *the number of holes injected from the collector to the base increases*. The collector hole flux is opposite to the flux of holes arriving from the emitter; therefore the two components subtract. As illustrated in Fig. 2.12(b), the net collector current decreases as V_{CB} increases (that is, as V_{EC} decreases) with a fixed value of I_B (which in turn reduces V_{EB}). This is more clearly seen by inspection of Eq. (2.32c).

2.4.3. I_{ECO} for the *pnp* and I_{CEO} for the *npn*

The saturation current from the emitter to the collector with the base open ($I_B = 0$) is defined as I_{CE0} on transistor data sheets. However, it should be defined for the *pnp* device as I_{ECO} since the current is from the E–C. This current is the collector current near the edge of cutoff, as illustrated in Fig. 2.12(b). For $I_C < I_{ECO}$, the base current must be negative and the device is generally into the cutoff region.

A physical explanation for why $I_{ECO} > I_{BCO}$ for the *pnp* is based on the "transistor action" of the electrons in the base being back injected into the forward biased emitter. Figure 2.13 illustrates the currents and particle fluxes of the common emitter *pnp* with the base open circuited. The minority electrons within one diffusion length (L_C) of the C–B junction are swept through the depletion region of the reverse biased C–B junction. When they reach the base they are majority carriers and easily push electrons to the forward biased B–E junction. At the B–E junction the electrons become the source of electrons to be injected into the emitter. Because of the p^+ doping of the emitter, this relatively small number of electrons, originating in the collector, forces the injection of large numbers of holes into the base (since $\gamma \rightarrow 1$). For the ideal device with no recombination in the base, these holes reach the B–C junction and are collected by the collector ($I_{Ep} = I_{Cp}$). The total collector current, as illustrated by Fig. 2.13, is

$$I_{ECO} = I_{Cn} + I_{Cp} \cong I_{BCO} + I_{Ep} \tag{2.41}$$

where I_{Cn} is approximately the reverse saturation current of the C–B junction with the emitter open, and I_{Cp} represents the injected holes from the emitter that reach the collector. Also note that in this particular case, with no generation or recombination in the base, $I_{BCO} = I_{En}$, the back injected electrons from base to emitter. From Eq. (2.39), repeated here for convenience,

$$\beta_{dc} \cong I_{Ep}/I_{En}$$

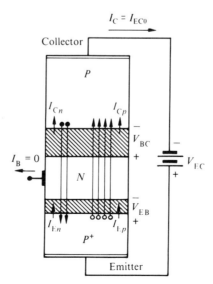

Fig. 2.13   The current I_{ECO} for a *pnp*.

Therefore Eq. (2.40) can be written as

$$I_{ECO} = I_{En} + \beta_{dc} I_{En} = I_{BCO}(\beta_{dc} + 1) \qquad (2.42)$$

To summarize, the collector current with the base open is much larger than just the reverse saturation (leakage) current of the reverse biased C–B junction because of the "transistor action" caused by the base current electrons forward biasing the E–B junction. Note the current gain of the device! This is an excellent example of the "transistor action" in the BJT device.

Connecting the external resistor across the E–B junction of Fig. 2.14 illustrates the case where I_C is made less than I_{ECO} by removing some of the electrons which originated at the C–B junction, thereby preventing some of them from reaching the E–B junction. Note that the electrons extracted from the base result in a negative base current. In the case shown the forward biased E–B junction causes current to flow in the resistor (R). The negative base current "steals" electrons from the base, thereby denying some back injected electrons to the emitter. With fewer back injected electrons the hole current from E–B is reduced, forcing I_C to be less than I_{ECO}. In the limit, as R is reduced to zero (and therefore $V_{EB} = 0$), I_C becomes approximately I_{BCO} and I_B approximately $-I_{BCO}$.

The *npn* transistor has a similar set of currents, I_{CEO} and I_{CBO}, which can be obtained in two ways. A simple way is to use the "complement" of the *pnp* by interchanging

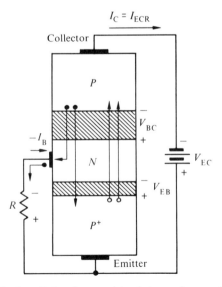

Fig. 2.14 $I_{BCO} \lessgtr I_{ECO}$ for a resistor between base and emitter.

holes for electrons and reversing each current direction and voltage polarity. The other way to generate equations for the *npn* is to derive Eqs. (2.41) and (2.42).

2.5 EBERS–MOLL EQUATIONS

Many of the computer aided circuit analysis (CAD) programs written to solve transistor circuit problems use the nonlinear Eqs. (2.27), (2.30), and (2.31) or Eqs. (2.32a) to (2.32c) to solve for the dc operating point variables. The computer must be used in their solution because of the nonlinear relationships between the currents and junction voltages. SPICE2 is a CAD program very often used for the analysis of bipolar (as well as junction field effect and MOSFET) transistor circuits which uses a form of these equations. We simplify the equations by collecting terms into only several constants.

The coefficients of Eq. (2.27) are given different names; for example, the first term is defined as

$$I_F = I_{F0}(e^{qV_{EB}/kT} - 1) = qA\left[\frac{D_E n_{E0}}{L_E} + \frac{D_B p_{B0}}{W}\right](e^{qV_{EB}/kT} - 1) \qquad (2.43)$$

where

$$I_{F0} = qAn_i^2\left[\frac{D_E}{L_E N_E} + \frac{D_B}{WN_B}\right]$$

and the second term as

$$\alpha_R I_R = \alpha_R I_{R0}(e^{qV_{CB}/kT} - 1) = \frac{qAD_B p_{B0}}{W}(e^{qV_{CB}/kT} - 1) \qquad (2.44)$$

Therefore the emitter current of Eq. (2.27) can be written as Eq. (2.45), which is much more compact than Eq. (2.27):

$$\boxed{I_E = I_F - \alpha_R I_R} \qquad (2.45)$$

Similarly for the collector current of Eq. (2.30), the second term is defined as

$$I_R = I_{R0}(e^{qV_{CB}/kT} - 1) = qA\left[\frac{D_C n_{C0}}{L_C} + \frac{D_B p_{B0}}{W}\right](e^{qV_{CB}/kT} - 1) \qquad (2.46)$$

where

$$I_{R0} = qAn_i^2 \left[\frac{D_C}{L_C N_C} + \frac{D_B}{W N_B} \right]$$

and the first term as

$$\alpha_F I_F = \alpha_F I_{F0}(e^{qV_{EB}/kT} - 1) = \frac{qAD_B}{W} p_{B0}(e^{qV_{EB}/kT} - 1) \tag{2.47}$$

Therefore, the collector current of Eq. (2.30) is written as

$$\boxed{I_C = \alpha_F I_F - I_R} \tag{2.48}$$

Since $I_B = I_E - I_C$, then Eq. (2.31) becomes

$$\boxed{I_B = (1 - \alpha_F)I_F + (1 - \alpha_R)I_R} \tag{2.49}$$

by subtracting Eq. (2.48) from Eq. (2.45). Equations (2.45), (2.48), and (2.49) are known as the *Ebers–Moll equations* for an ideal *pnp*. Figure 2.15 shows the Ebers–

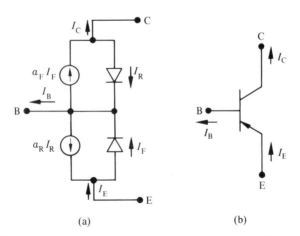

Fig. 2.15   (a) Ebers–Moll equivalent circuits: *pnp*; (b) symbol.

Moll equivalent circuit. This circuit is just a restatement of the equations. Note the similarity of I_F and I_R to the E–B and C–B junction diode formulas. The current sources $\alpha_F I_F$ and $\alpha_R I_R$ illustrate the interaction terms of the two junctions due to the narrow base region.

The coefficients in front of the exponential terms on the right sides of Eqs. (2.44) and (2.47) are identical; therefore

$$\boxed{\alpha_F I_{F0} = \alpha_R I_{R0} = I_S} \tag{2.50}$$

where I_S is defined by Eq. (2.50). The term I_S is one of the required values for SPICE2. We can see that if

$$\beta_F = \frac{\alpha_F}{1 - \alpha_F}$$

then only three numbers are necessary for the Ebers–Moll equations to be completely specified: β_F, β_R, and I_S. All the other parameters can be calculated from these three. It must also be noted that the Ebers–Moll equations can be applied to all regions of operation of the transistor.

The direct connection between doping densities, base width, lifetimes, etc., and the Ebers–Moll equations makes them particularly attractive in integrated circuit analysis. Note that the complete set of input–output V–I characteristics can be calculated from these equations. Also note that β_F and β_R are defined for all regions of operation and that in the active region $\beta_F \cong \beta_{dc}$.

Because it is the "complement" of a *pnp*, the *npn* has a set of Ebers–Moll equations similar to Eqs. (2.45), (2.48), and (2.49) and an equivalent circuit similar to Fig. 2.15 with the diodes and currents all reversed.

The Ebers–Moll equations can also be obtained for the nonideal case of recombination in the base by defining the appropriate coefficients in front of the exponential terms (I_{R0} and I_{F0}).

> **SEE EXERCISE 2.2–APPENDIX A**

> **SEE EXERCISE 2.3–APPENDIX A**

2.6 SUMMARY

A "game plan" for arriving at the solution for a *pnp* transistor was developed to guide the reader through to a derivation of the terminal currents. The terminal currents were obtained by evaluating the minority carrier diffusion currents at the edge of each depletion region. Because no g–r (generation–recombination) occurs in the depletion region, the currents must be constant throughout. Solving the minority carrier diffusion equations in the emitter, base, and collector with the appropriate boundary conditions resulted in the carrier distributions in each bulk region.

The *ideal bipolar transistor* was defined as a device having no recombination or generation in the bulk base region; that is, the holes that enter the base from the emitter must leave via the collector. The "game plan" as applied to the ideal *pnp* resulted in equations for I_E, I_C, and I_B, which were functions of V_{CB} and V_{EB}. These three equations represent the dc nonlinear model for the device in all four regions of operation; that is, active, saturation, cutoff, and inverted operation.

Carrier flux and electron and hole current components were obtained for the four regions of operation. The current equations were simplified in each region to the major components. If the reader is to fully understand the bipolar junction transistor, the carrier flux diagrams are very important.

The common base and common emitter $V–I$ characteristics were examined for the ideal device with alpha, beta, and the emitter-injection efficiency derived in terms of the material parameters. Two currents, I_{BCO} and I_{ECO}, were examined and their different magnitudes explained in terms of the transistor action. The chapter concluded with a discussion of the Ebers–Moll equations.

PROBLEMS

2.1 Derive a set of equations similar to Eqs. (2.32a), (b), and (c) for the ideal *npn* bipolar transistor starting with the solutions to the minority carrier diffusion equations.

2.2 Consider an ideal *pnp* bipolar transistor device operating with the following parameters:

Emitter	Base	Collector
$n_{E0} = 2.56 \times 10^2/\text{cm}^3$	$p_{B0} = 6.39 \times 10^3/\text{cm}^3$	$n_{C0} = 4.92 \times 10^5/\text{cm}^5$
$L_E = 22.8 \times 10^{-4}$ cm	$L_B = 46.9 \times 10^{-4}$ cm	$L_C = 39.5 \times 10^{-4}$ cm
$D_E = 5.18\ \text{cm}^2/\text{s}$	$D_B = 22\ \text{cm}^2/\text{s}$	$D_C = 15.6\ \text{cm}^2/\text{s}$
	$W = 4 \times 10^{-4}$ cm $= 4\ \mu\text{m}$	
$N_{AE} \cong 3.9 \times 10^{17}/\text{cm}^3$	$N_{DB} \cong 1.57 \times 10^{16}/\text{cm}^3$	$N_{AC} \cong 2 \times 10^{14}/\text{cm}^3$

Let $A = 1.265 \times 10^{-4}\ \text{cm}^2$, $n_i = 10^{+10}/\text{cm}^3$, and $kT = 0.026$ eV. Calculate at $V_{EB} = 0.67235$ and $V_{CB} = -1$, assuming that $W = 4\ \mu\text{m}$ at these voltages:

(a) The current components I_{En}, I_{Ep}, and I_E

(b) The current components I_{Cp}, I_{Cn}, and I_C

(c) I_{B1} and I_{B3}; also, what % of the total I_B is I_{B3}?

(d) How do the exponential terms containing V_{CB} compare in size with those containing V_{EB}?

(e) What is the emitter injection efficiency, α_{dc}, and β_{dc}?

[*Suggestion:* Make a copy of your calculations, as they will be used in other problems.]

2.3 Using the concept of "complementary" devices,

(a) Sketch the minority carrier concentrations for an ideal n^+pn device in the cutoff, saturation, and inverted active regions of operation.

(b) Make a sketch similar to Fig. 1.9 for each region of operation in part (a).

2.4 Most modern BJTs have an ion implanted and diffused base which has a Gaussian impurity profile. It can be shown that as far as the electric field is concerned this impurity profile can be approximated as an exponential in the bulk base region. If the p^+np base doping profile is approximated by

$$N_{DB} = N_D(0)e^{-ax/W}, \quad \text{where } a = \ln\left[\frac{N(0)}{N(W)}\right]$$

(a) Show that away from the edges of the base region the electric field is constant. Assume the device to be in thermal equilibrium. [*Hint:* $\mathscr{E} \propto a/W$.]

(b) Assume no recombination in the base and derive a formula for $p_B(x)$; let $p_B(W) = 0$ and $I_E \cong I_{Ep}$, the hole current in the base.

(c) Calculate and plot $p_B(x)/(I_{Ep}W/qAD_Ba)$ versus x/W for $a = 5$ and $a = 9$.

(d) What can be surmised about the drift and diffusion components of hole current at 0.2 W, 0.5 W, and 0.9 W for $a = 9$?

(e) Explain how this "graded base" will affect the chance of any recombination in the base.

2.5 In Problem 2.2 the base width is halved to 2 μm:

(a) Calculate the new values of β_{dc} and α_{dc}.

(b) Calculate the new value of I_{Ep} and compare it to the original value.

(c) How has this change affected the emitter injection efficiency?

2.6 Calculate the Ebers–Moll coefficients for the ideal *pnp* device of Problem 2.2, and examine the output I–V characteristic of the common base as the device is operated from forward active into deep saturation with $I_E = 200$ μA by plotting I_C versus V_{BC} for several values of V_{BC}. At each voltage, calculate V_{EB}, V_{EC}, and the current components I_{Cp} and I'_{Cp}. V_{BC} equals

(a) 1

(b) 0

(c) -0.10

(d) -0.60

2.7 Derive an equation for V_{EB} in terms of the Ebers–Moll coefficients to be used for plotting the common emitter I_C vs V_{EC}. Note that I_B is a constant and V_{EC} is a variable used in plotting of the output characteristic.

2.8 Calculate the Ebers–Moll coefficients for the ideal *pnp* device of Problem 2.2, and examine the output *I–V* characteristic of the common emitter as the device is operated from active into deep saturation with $I_B = 2$ μA by plotting I_C versus V_{EC} for several values of V_{CB}:

(a) $V_{CB} = -1$

(b) $V_{CB} = 0$ (edge of saturation active region)

(c) $V_{CB} = +0.45$. What can be said about the amount of forward bias of V_{CB} necessary to significantly reduce I_C from its active region value?

2.9 In many modern BJT devices, in order to obtain very high speed it is necessary to have very thin emitter bulk regions (such a device is called a "shallow" emitter). If the device is ideal and $L_E \gg W_E$ for the *pnp* BJT in Fig. P2.9 and $\Delta n(x'' = W_E) = 0$ due to the metal contact,

(a) Sketch the minority carrier profiles for active region operation and derive an equation for the emitter current component I_{En}.

(b) Derive an equation for the total emitter current.

(c) Obtain the emitter injection efficiency.

(d) Explain what happens to the beta of the device as W_E is made smaller.

2.10 With reference to Problem 2.9, recent work has shown that a polysilicon emitter (or, more correctly, a polysilicon contacted emitter) can be used to change the minority carrier concentration profile in the shallow emitter. As a result, a very fast BJT is produced without reducing the beta significantly. In Fig. P2.10 let $L_E \gg W_E$ and $\Delta n(x'' = W_E) = K$, and assume that the polysilicon is very thick compared with the minority carrier diffusion length in that region:

(a) Sketch the minority carrier profiles for active region operation and derive an equation for the emitter current component I_{En}.

(b) Derive an equation for the total emitter current.

(c) Obtain the emitter injection efficiency.

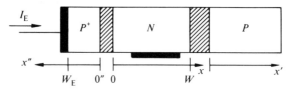

Fig. P2.9

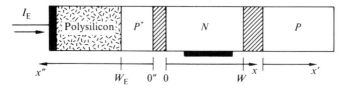

Fig. P2.10

3 / Deviations from the Ideal Transistor

The ideal bipolar transistor presented in Chapter 2 more than adequately describes most real devices in all their regions of normal operation. However, at very small currents and voltages or at large currents and voltages, the V–I characteristics of a real device, fabricated in silicon, deviate from those of the ideal device. The purpose of this chapter is to discuss the reasons for these deviations from the ideal bipolar transistor.

The first deviation from ideal is that recombination is allowed to occur in the quasi-neutral base region. We first derive a very general set of equations that require hyperbolic functions, but that apply to all possible cases of base width and region of operation. An approximation to this case is obtained by allowing for very small base recombination and yet letting the base minority carrier profile be a straight line. This can be called the "quasi-ideal BJT." Here we clearly see the increased base current which accounts for minority carrier recombination in the base.

Comparisons between the ideal and observed V–I characteristics show that the basic shapes of the characteristics are about the same for the real and ideal devices. However, theoretically one predicts a single input characteristic for output voltages whose magnitudes are greater than a few kT/q volts. In reality, the input characteristics are seen to change somewhat as the output voltage is made more negative; that is, the C–B becomes more reverse biased. Likewise, the theoretical output characteristics are flat over most of the output voltage range. The observed output characteristics, on the other hand, slant upward with increasing magnitude of the output voltage.

Generation–recombination in the E–B and C–B depletion regions, junction avalanche voltage breakdown, series resistance, and high-level injection phenomena are also deviations from the ideal. These are similar to those in the p-n diode. Several new effects are punch-through, current crowding, and collector-to-emitter voltage breakdown.

3.1 RECOMBINATION IN THE BASE

The minority carrier concentration in the base for a *pnp* device, $p_B(x)$, is altered from a straight line to a hyperbolic function when significant recombination is permitted in the base region. A complete derivation is very similar to that of the medium base diode (see problems in Chapter 3, Vol. II) with a different function for $p_B(x)$ due to the boundary conditions at the C–B junction.

The derivation begins with the minority carrier diffusion equation in the base, Eq. (3.1a), which has Eq. (3.1b) as a solution:

$$\frac{d^2 \Delta p_B(x)}{dx^2} = \frac{\Delta p_B(x)}{D_B \tau_B} = \frac{\Delta p_B(x)}{L_B^2} \tag{3.1a}$$

$$\Delta p_B(x) = C_1 e^{-x/L_B} + C_2 e^{x/L_B} \tag{3.1b}$$

The boundary conditions are determined by the junction voltages V_{EB} and V_{CB}, Eqs. (2.9) and (2.10). At $x = 0$,

$$\Delta p_B(0) = C_1 + C_2 = p_{B0}(e^{qV_{EB}/kT} - 1) \tag{3.1c}$$

and at $x = W$,

$$\Delta p_B(W) = C_1 e^{-W/L_B} + C_2 e^{W/L_B} = p_{B0}(e^{qV_{CB}/kT} - 1) \tag{3.1d}$$

Equations (3.1c) and (3.1d) have two unknowns and must be solved simultaneously for the constants C_1 and C_2. Equation (3.1c) can be solved for C_1 and the result substituted into Eq. (3.1d). Now C_2 can be solved for in terms of the boundary conditions $\Delta p_B(0)$ and $\Delta p_B(W)$. After some algebraic manipulation, we obtain

$$C_1 = \frac{\Delta p_B(0)e^{W/L_B} - \Delta p_B(W)}{e^{W/L_B} - e^{-W/L_B}} \tag{3.2a}$$

Now either Eq. (3.1c) or Eq. (3.1d) can be used to solve for C_2 in terms of the boundary conditions; i.e., after some algebraic manipulation we get

$$C_2 = \frac{\Delta p_B(W) - \Delta p_B(0)e^{-W/L_B}}{e^{W/L_B} - e^{-W/L_B}} \tag{3.2b}$$

Placing C_1 and C_2 back into Eq. (3.1b) we get Eq. (3.3):

$$\Delta p_B(x) = \left[\frac{\Delta p_B(0)e^{W/L_B} - \Delta p_B(W)}{e^{W/L_B} - e^{-W/L_B}}e^{-x/L_B}\right] + \left[\frac{\Delta p_B(W) - \Delta p_B(0)e^{-W/L_B}}{e^{W/L_B} - e^{-W/L_B}}\right]e^{+x/L_B}$$

$$\tag{3.3}$$

The hyperbolic functions are defined as

$$\sinh(x) = \frac{e^x - e^{-x}}{2} \quad \text{and} \quad \cosh(x) = \frac{e^x + e^{-x}}{2}$$

Note that the denominator of Eq. (3.3) is $2 \sinh(W/L_B)$. Expanding Eq. (3.3) yields

$$\Delta p_B(x) = \frac{1}{2 \sinh\left(\dfrac{W}{L_B}\right)} [\Delta p_B(0)e^{W/L_B}e^{-x/L_B} - \Delta p_B(W)e^{-x/L_B} + \Delta p_B(W)e^{x/L_B}$$

$$- \Delta p_B(0)e^{-W/L_B}e^{x/L_B}]$$

$$\Delta p_B(x) = \frac{1}{2 \sinh\left(\dfrac{W}{L_B}\right)} [\Delta p_B(0)(e^{(W-x)/L_B} - e^{-(W-x)/L_B}) + \Delta p_B(W)(e^{x/L_B} - e^{-x/L_B})]$$

which can be simplified again using the definition for $\sinh(x)$ as shown in Eq. (3.4):

$$\Delta p_B(x) = \frac{1}{2 \sinh\left(\dfrac{W}{L_B}\right)} \left[\Delta p_B(0)2 \sinh\left(\frac{W - x}{L_B}\right) + \Delta p_B(W)2 \sinh\left(\frac{x}{L_B}\right) \right] \qquad (3.4)$$

Cancelling the "twos" and then replacing the boundary conditions $\Delta p_B(0)$ and $\Delta p_B(W)$ with their equivalent voltage representations results in Eqs. (3.5a) and (3.5b):

$$\boxed{\Delta p_B(x) = \frac{1}{\sinh\left(\dfrac{W}{L_B}\right)} \left[\Delta p_B(0) \sinh\left(\frac{W - x}{L_B}\right) + \Delta p_B(W) \sinh\left(\frac{x}{L_B}\right) \right]}$$

$$(3.5a)$$

$$\Delta p_B(x) = p_{B0}(e^{qV_{EB}/kT} - 1)\frac{\sinh\left(\dfrac{W - x}{L_B}\right)}{\sinh\left(\dfrac{W}{L_B}\right)} + p_{B0}(e^{qV_{CB}/kT} - 1)\frac{\sinh\left(\dfrac{x}{L_B}\right)}{\sinh\left(\dfrac{W}{L_B}\right)}$$

$$(3.5b)$$

Figure 2.6 illustrates the hole concentrations for the active, saturation, and cutoff regions as V_{EB} and V_{CB} are changed appropriately.

SEE EXERCISE 3.1–APPENDIX A

To obtain the emitter current we go back to our "game plan" and evaluate the electron and hole currents at each edge of the E–B depletion region. Note that for the *pnp* device the electron current I_{En} will be the same as before since only $p_B(x)$ was changed from the ideal transistor solution.

$$I_{Ep} = -qAD_B \frac{d\,\Delta p_B(x)}{dx}\bigg|_{x=0} \tag{3.6}$$

Therefore I_{Ep} is obtained from Eq. (3.5a) or (b). Since the differential of $\sinh(x)$ equals $\cosh(x)$, then, from Eq. (3.6),

$$I_{Ep} = \frac{-qAD_B}{\sinh(W/L_B)}\left[\Delta p_B(0)\cosh\left(\frac{W-x}{L_B}\right)\left[\frac{-1}{L_B}\right] + \Delta p_B(W)\cosh\left(\frac{x}{L_B}\right)\left[\frac{+1}{L_B}\right]\right]_{x=0} \tag{3.7a}$$

$$I_{Ep} = qA\frac{D_B}{L_B}p_{B0}\left[\frac{\cosh\left(\dfrac{W}{L_B}\right)}{\sinh\left(\dfrac{W}{L_B}\right)}(e^{qV_{EB}/kT}-1) - \frac{(e^{qV_{CB}/kT}-1)}{\sinh\left(\dfrac{W}{L_B}\right)}\right] \tag{3.7b}$$

With the aid of Eq. (2.24) for I_{En} the total emitter current is determined to be

$$I_E = qA\left[\frac{D_E}{L_E}n_{E0} + \frac{D_B}{L_B}p_{B0}\coth\left(\frac{W}{L_B}\right)\right](e^{qV_{EB}/kT}-1) - \left[\frac{qAD_B p_{B0}}{L_B\sinh\left(\dfrac{W}{L_B}\right)}\right](e^{qV_{CB}/kT}-1) \tag{3.8a}$$

or

$$I_E = qAn_i^2\left[\frac{D_E}{L_E N_E} + \frac{D_B}{L_B N_B}\coth\left(\frac{W}{L_B}\right)\right](e^{qV_{EB}/kT}-1) - qAn_i^2\left[\frac{D_B}{L_B N_B}\frac{1}{\sinh\left(\dfrac{W}{L_B}\right)}\right]$$

$$\cdot (e^{qV_{CB}/kT}-1) \tag{3.8b}$$

Note that the forms of Eq. (3.8a) and (b) are very similar to the ideal case, Eq. (3.32a) and (b).

The collector current can be derived in a similar manner by evaluating the following equation, which is made up of I_{Cn} and I_{Cp}, respectively:

$$I_C = -qAD_B\frac{d\,\Delta p_B}{dx}\bigg|_{x=W} + qAD_C\frac{d\,\Delta n_C}{dx'}\bigg|_{x'=0}$$

The result is

$$I_C = qA\left[\frac{D_B}{L_B}p_{B0}\frac{1}{\sinh\left(\dfrac{W}{L_B}\right)}\right](e^{qV_{EB}/kT}-1)$$

$$- qA\left[\frac{D_C}{L_C}n_{C0} + \frac{D_B p_{B0}}{L_B}\coth\left(\frac{W}{L_B}\right)\right](e^{qV_{CB}/kT}-1) \tag{3.9a}$$

or

$$I_C = qAn_i^2 \left[\frac{D_B}{L_B N_B} \frac{1}{\sinh\left(\dfrac{W}{L_B}\right)} \right] (e^{qV_{EB}/kT} - 1) - qAn_i^2 \left[\frac{D_C}{L_C N_C} + \frac{D_B}{L_B N_B} \coth\left(\frac{W}{L_B}\right) \right]$$

$$\cdot (e^{qV_{CB}/kT} - 1) \tag{3.9b}$$

Since $I_B = I_E - I_B$, we can subtract Eq. (3.9b) from Eq. (3.8b):

$$\boxed{\begin{aligned} I_B = {} & qAn_i^2 \left[\frac{D_E}{L_E N_E} + \frac{D_B}{L_B N_B} \left(\coth\left(\frac{W}{L_B}\right) - \frac{1}{\sinh\left(\dfrac{W}{L_B}\right)} \right) \right] (e^{qV_{EB}/kT} - 1) \\ & + qn_i^2 \left[\frac{D_C}{L_C N_C} + \frac{D_B}{L_B N_B} \left(\coth\left(\frac{W}{L_B}\right) - \frac{1}{\sinh\left(\dfrac{W}{L_B}\right)} \right) \right] (e^{qV_{CB}/kT} - 1) \end{aligned}}$$

$$\tag{3.10}$$

Note that as $x \ll 1$, $\cosh(x) \cong 1$ and $\sinh(x) \cong x$ so that $\coth(x) \cong 1/x$. Therefore, if $W \ll L_B$ in Eq. (3.10),

$$\coth\left(\frac{W}{L_B}\right) - \frac{1}{\sinh\left(\dfrac{W}{L_B}\right)} \cong \frac{L_B}{W} - \frac{1}{\dfrac{W}{L_B}} \cong 0$$

and we approach the ideal BJT base current equation, Eq. (2.32c). The hyperbolic term obviously represents the base recombination which was labeled I_{B2} in Fig. 3.1(a). As illustrated, additional electrons must enter the base to neutralize the holes recombining in the base width W. Due to the minority carrier recombination, the base current has increased ($I_{B2} > 0$), and the base transport factor, α_T, has decreased since $I_{Cp} < I_{Ep}$. Also, α_{dc} has decreased, and hence β_{dc} has also decreased.

SEE EXERCISE 3.2–APPENDIX A

3.1.1 Quasi-ideal

A compromise between the ideal device (no recombination in W) and the general case, which results in rather abstract mathematical functions, is the quasi-ideal device. Since for most modern devices $W < L_B$, and yet recombination may be significant, we want

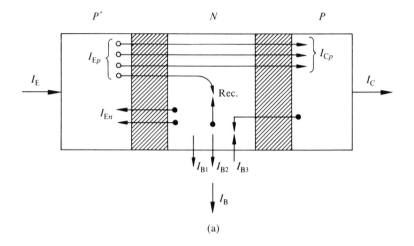

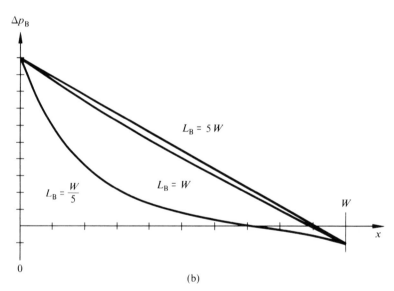

Fig. 3.1 p^+np BJT in the active region with base recombination: (a) carrier flux; (b) hole concentration for values of L_B.

to simplify the general hyperbolic current equations. Two methods are presented. We define a *quasi-ideal device* as having a straight line minority carrier concentration distribution in the base, as indicated by Eq. (2.19) and plotted in Fig. 2.5; yet it has recombination in the base ($I_{B2} \neq 0$). When $\Delta p_B(x)$ is plotted for a device with W/L_B of about one-half, the curve appears to approximate a straight line. Only when it is compared with a truly straight line is the difference readily apparent. See Fig. 3.1(b) for several examples. Most modern devices have W/L_B of one-tenth or less, and therefore a straight line approximation, in addition to recombination, is reasonable.

The recombination current component of the base current, I_{B2}, is obtained by determining the total injected charge in the base which recombines on the average of every τ_B seconds, which will be the current due to recombination. With the aid of Eq. (2.19) we find the total injected charge by integrating $\Delta p_B(x)$ and multiplying by the area:

$$I_{B2} = \frac{\Delta Q_B}{\tau_B} = \frac{qA}{\tau_B} \int_0^W \Delta p_B(x)\, dx = \frac{qA}{\tau_B} \int_0^W \left[\Delta p_B(0) - \left(\frac{\Delta p_B(0) - \Delta p_B(W)}{W} \right) x \right] dx$$

$$(3.11a)$$

$$I_{B2} = \frac{qA}{2\tau_B} [\Delta p_B(0) + \Delta p_B(W)]W = \frac{qAW}{2\tau_B} p_{B0}[(e^{qV_{EB}/kT} - 1) + (e^{qV_{CB}/kT} - 1)]$$

$$(3.11b)$$

Note that the above equation is valid for all regions of operation and is simply added to Eq. (2.31), yielding a base current increased to

$$I_B = qA \left[\frac{D_E}{L_E} n_{E0} + \frac{W}{2\tau_B} p_{B0} \right] (e^{qV_{EB}/kT} - 1) + qA \left[\frac{D_C}{L_C} n_{C0} + \frac{W}{2\tau_B} p_{B0} \right] (e^{qV_{CB}/kT} - 1)$$

$$(3.12a)$$

Since $p_{B0} = n_i^2/N_B$ and $L_B^2 = \tau_B D_B$, Eq. (3.12a) can be written in a different form. The second term inside the bracket is

$$\frac{W}{2\tau_B} p_{B0} = \frac{W}{2L_B^2/D_B} \times \frac{n_i^2}{N_B} = \frac{W n_i^2 D_B}{2L_B^2 N_B}$$

and therefore

$$I_B = qA n_i^2 \left[\frac{D_E}{L_E N_E} + \frac{D_B}{L_B N_B} \times \frac{W}{2L_B} \right] (e^{qV_{EB}/kT} - 1)$$

$$+ qA n_i^2 \left[\frac{D_C}{L_C N_C} + \frac{D_B}{L_B N_B} \times \frac{W}{2L_B} \right] (e^{qV_{CB}/kT} - 1) \qquad (3.12b)$$

Note that as $W/L_B \to 0$, the recombination base current $I_{B2} \to 0$. For typical devices, the first term of Eq. (3.12a) may have $n_{E0} \sim 10^2$, $p_{B0} \sim 10^4$, $D_E \sim 10$, $L_E \sim 25$ μm, $W \sim 1.5$ μm, and $\tau_B \sim 1$ μs, which yields

$$D_E n_{E0}/L_E \sim (10 \times 10^2)/25 \ \mu\text{m} = 4 \times 10^5$$

while $Wp_{B0}/2\tau_B \sim 1.510^{-4} \times 10^4/2 \times 10^{-6} = 7.5 \times 10^5$, which indicates that I_{B1} and I_{B2} could be of comparable size.

The emitter current in the active region is not significantly affected by recombination in the base for two reasons. First, I_{B2}, like I_{B1}, is much smaller than the injected hole current I_{Ep}, and $I_E \cong I_{Ep}$. Secondly, I_{Ep} is determined by the slope of $p_B(x)$ at $x = 0$, which does not change much from the ideal straight line case. We therefore do not change I_E from the ideal case.

The collector hole current I_{Cp} is reduced because fewer holes arrive at the collector. Holes lost in recombination with electrons in the bulk base region create I_{B2} and therefore reduce I_C by that amount:

$$I_C = I_{C_{ideal}} - I_{B2} \tag{3.13}$$

From Eq. (2.30),

$$I_C = qA\left[\frac{D_B p_{B0}}{W} - \frac{Wp_{B0}}{2\tau_B}\right](e^{qV_{EB}/kT} - 1) - qA\left[\frac{D_C n_{C0}}{L_C} + \frac{D_B p_{B0}}{W} - \frac{p_{B0}W}{2\tau_B}\right](e^{qV_{CB}/kT} - 1) \tag{3.14}$$

Again note that

$$D_B/W \sim \frac{20}{10^{-4}} = 2 \times 10^5 \quad \text{and} \quad \frac{W}{2\tau_B} \sim \frac{10^{-4}}{2 \times 10^{-6}} = 50$$

and I_{B2} will not change I_C significantly in most cases. Therefore, in the quasi-ideal device in the active region,

$$
\begin{array}{ll}
I_E \cong I_{E_{ideal}} & \text{as in Eq. (2.27)} \\
I_C \cong I_{C_{ideal}} & \text{as in Eq. (2.30)} \\
I_B \cong I_{B_{ideal}} + I_{B2} & \text{as in Eq. (3.12)}
\end{array}
$$

The quasi-ideal bipolar transistor is a good compromise between the simplicity of tracing the current components and the accuracy of the complete hyperbolic relationships.

3.1.2 Complete Quasi-ideal Transistor

A more complete approach to the nonideal device, in which there is some recombination in the base, is to assume $W/L_B \ll 1$ in Eqs. (3.8), (3.9), and (3.10). The basis for this argument is that in most modern devices $W \cong 0.5$ μm while $L_B \cong 20$ μm. A series expansion of $\cosh(x)$ and $\sinh(x)$ for $x \ll 1$ is

$$\cosh(x) \cong 1 + (1/2)x^2 + \cdots$$

$$\sinh(x) \cong x + \cdots$$

$$\coth(x) \cong \frac{x + (1/2)x^2 \cdots}{x \cdots} \cong \frac{1}{x} + \frac{x}{2} \cdots \cong \frac{1}{\sinh(x)} + \frac{x}{2}$$

In Eq. (3.10),

$$\coth(W/L_B) - \frac{1}{\sinh(W/L_B)} \cong \frac{1}{W/L_B} + \frac{W/L_B}{2} - \frac{1}{W/L_B} \cong \frac{W}{2L_B}$$

and

$$I_B = qAn_i^2 \left[\frac{D_E}{L_E N_E} + \frac{D_B}{L_B N_B} \times \frac{W}{2L_B} \right] (e^{qV_{EB}/kT} - 1)$$
$$+ qAn_i^2 \left[\frac{D_C}{L_C N_C} + \frac{D_B}{L_B N_B} \times \frac{W}{2L_B} \right] (e^{qV_{CB}/kT} - 1) \qquad (3.15)$$

which is identical to Eq. (3.12b).

Applying the hyperbolic approximation to I_E, Eq. (3.8b) yields

$$I_E = qAn_i^2 \left[\frac{D_E}{L_E N_E} + \frac{D_B}{L_B N_B} \left(\frac{L_B}{W} + \frac{W}{2L_B} \right) \right] (e^{qV_{EB}/kT} - 1)$$
$$- qAn_i^2 \left[\frac{D_B}{L_B N_B} \times \frac{L_B}{W} \right] (e^{qV_{CB}/kT} - 1) \qquad (3.16)$$

Similarly for I_C

$$I_C = qAn_i^2 \left[\frac{D_B}{L_B N_B} \left(\frac{L_B}{W} \right) \right] (e^{qV_{EB}/kT} - 1)$$
$$- qAn_i^2 \left[\frac{D_C}{L_C N_C} + \frac{D_B}{L_B N_B} \left(\frac{L_B}{W} + \frac{W}{2L_B} \right) \right] (e^{qV_{CB}/kT} - 1) \qquad (3.17)$$

Note that $L_B/W \gg W/2L_B$ and therefore Eqs. (3.16) and (3.17) reduce to the quasi-ideal equations of the previous section.

3.2 BASE WIDTH MODULATION

In drawing the theoretical ideal characteristics of Chapter 2 we implicitly assumed α_{dc}, β_{dc}, I_{BCO}, and I_{ECO} to be constants independent of the applied voltages. All of these quantities are functions of W. We have assumed that W, the quasi-neutral width of the base, is independent of bias. In truth, W is not a constant independent of voltage. Changing V_{EB} and/or V_{CB} changes the depletion widths about the E–B and/or C–B junctions, thereby changing W as illustrated in Fig. 3.2(a). The changes in the depleted portions of the base are especially significant because of the narrow physical extent of the base. A small change of only 0.05 μm may be a significant percentage of the base width W.

The above effect, known as *base width modulation* or the *Early effect* (named after the engineer who first explained the phenomenon), can now be used to explain the observed variation in the input characteristics with output voltage changes, and the sloping portions of the active region output characteristics as illustrated in Fig. 3.3.

Comparisons between the ideal and observed V–I characteristics are shown in Figs. 3.3 and 3.4. Figure 3.3 displays common base characteristics while Fig. 3.4 displays common emitter characteristics. As can be seen from these figures, the basic shapes of the characteristics are about the same for the fabricated and ideal devices. However, theoretically one predicts a single input characteristic for output voltages whose magnitudes are greater than a few kT/q volts. In reality, the input characteristics are seen to change somewhat as the output voltage (V_{CB}) is made more negative; that is, the C–B becomes more reverse biased. Likewise, the theoretical output characteristics are flat over most of the output voltage range. The observed output characteristics, on the other hand, slant upward with increasing magnitude of the output voltage.

As seen in Eq. (2.27), repeated here as Eq. (3.18), which describes the common base input characteristics, I_E is approximately proportional to $1/W$ and will therefore increase as W is decreased due to a more negative value of V_{CB}.

$$I_E = qA\left[\frac{D_E n_{E0}}{L_E} + \frac{D_B p_{B0}}{W}\right](e^{qV_{EB}/kT} - 1) - \frac{qAD_B}{W}p_{B0}(e^{qV_{CB}/kT} - 1) \qquad (3.18)$$

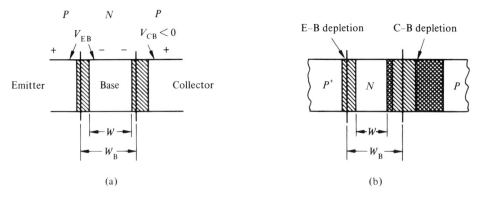

(a) (b)

Fig. 3.2 Base width modulation effects as V_{CB} becomes more negative: (a) initial; (b) larger reverse bias of C–B junction.

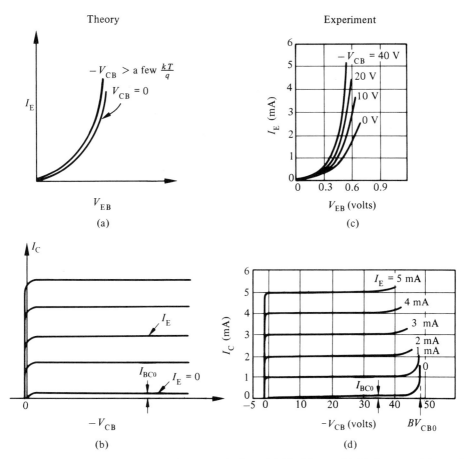

Fig. 3.3 Ideal versus real common base *pnp* devices: (a) ideal input; (b) ideal output; (c) actual input (experimental); (d) actual output (experimental).

For $V_{CB} < -0.1$ volt and $V_{EB} > 0.1$ volt, Eq. (3.18) can be approximated by

$$I_E \cong qA\left[\frac{D_E n_{E0}}{L_E} + \frac{D_B p_{B0}}{W}\right] e^{qV_{EB}/kT} \qquad (3.19)$$

Also remember that $p_{B0} \gg n_{E0}$ because the p^+ emitter is doped much greater than the base; hence

$$I_E \cong qA\frac{D_B p_{B0}}{W} e^{qV_{EB}/kT} \qquad (3.20)$$

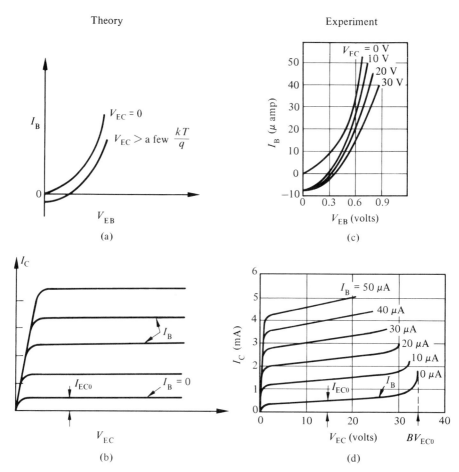

Fig. 3.4 Ideal versus real common emitter *pnp* devices: (a) ideal input; (b) ideal output; (c) actual input (experimental); (d) actual output (experimental).

The base width W clearly decreases as V_{CB} becomes more negative, as shown in Fig. 3.2(b), and hence I_E will be greater for a fixed V_{EB} as illustrated in Fig. 3.3(c).

The common base output characteristics of Fig. 3.3(d) show little base width modulation effects. Inspection of the active region shows only a very slight increase in I_C as V_{CB} changes for a fixed I_E. This phenomenon is somewhat self compensating because of the fixed emitter current condition. As W becomes smaller, I_E wants to increase; therefore V_{EB} decreases to keep I_E fixed. The result is fewer holes injected into the base; they are the primary source of carriers for I_C; that is, I_C will not increase significantly.

What slope exists in I_C versus V_{BC} is due to three possible phenomena. The first is a surface current around the reverse biased C–B junction, which in effect adds a small conductance in parallel with the junction. The second phenomenon occurs as W shrinks. With a smaller W there is even less of a chance that holes will recombine in the base, thereby increasing the number of injected holes that reach the collector; that is, there is a very slight increase in I_C. This second phenomenon will occur only in devices where significant recombination is present in the base. The third phenomenon is similar to the one observed in the reverse biased junction diode where generation in the depletion region increases the reverse saturation current. As V_{CB} becomes more negative, the C–B depletion region widens, allowing a larger generation current and thus an increase in I_C. This is observed only for small values of I_E.

Figures 3.2(c) and (d) present the common emitter input and output characteristics that show larger effects due to base width modulation. The Early effect on the actual device output characteristic shows a finite slope throughout the active region I_C versus V_{CE} plot (with a fixed I_B). Equation (2.31), for the ideal base current, shows that if I_B is to be a constant in the active region, V_{EB} must be constant, since by neglecting the last term (because V_{CB} is negative) one obtains

$$I_B \cong \frac{qAD_E}{L_E} n_{E0} e^{qV_{EB}/kT} + I_{B2} \tag{3.21}$$

Similarly, the collector current of Eq. (2.30), in the active region where $V_{EB} > 0.1$ volt, can be approximated by

$$I_C \cong \frac{qAD_B}{W} p_{B0} e^{qV_{EB}/kT} \tag{3.22}$$

For an increase in V_{EC} (V_{EB} constant), V_{CB} becomes more negative and the base width W decreases, causing an increase in I_C. Also note that, because $N_C \ll N_B$, most of the C–B depletion region is into the collector.

Another important phenomenon contributing to the slope of the common emitter output characteristics is the generation current produced in the C–B depletion region. Electrons and holes are generated, each contributing to an increase in I_C as the C–B depletion region increases. However, the largest increase is a result of the generated electrons drifting into the base, where they become majority carriers. These electrons add directly to the back injected electrons at the E–B junction, which forces a much larger increase in the injected holes from the emitter, which in turn diffuse to the collector junction and increase I_C significantly. Note that this device phenomenon is very similar to the I_{BC0} and I_{EC0} relationship in that "transistor action" has increased the effect. Figure 3.4(d), for $I_B = 0$, shows the effect of the C–B depletion width increase (more carrier generation) as it increases I_{EC0} with V_{EC} increasing.

3.3 PUNCH-THROUGH AND AVALANCHING

3.3.1 Punch-through

The term *punch-through* (or *reach-through*) refers to the physical situation where base width modulation has resulted in $W \to 0$; that is, the base is said to be punched-through when the E–B and C–B depletion regions touch each other inside the base, as illustrated in Fig. 3.5. Once punch-through has occurred, the emitter-base and collector-base junctions become electrostatically coupled. An increase in $-V_{CB}$ beyond the punch-through point causes a lowering of the E–B junction potential hill, as illustrated in Fig. 3.6, allowing a large (exponential) injection of holes from the emitter directly to the collector. Only a slight increase in V_{BC} is necessary for a large increase in I_C. The effect of punch-through on the output characteristics may be seen in Figs. 3.3(d) and 3.4(d) as the rapidly increasing I_C at large values of $-V_{CB}$ and V_{EC}, respectively. We say "may" because the $-V_{CB}$ required for punch-through may be greater than that required for another phenomenon giving rise to a similar high-current effect, namely *avalanching*. The high-current regions are caused by either avalanching or punch-through, whichever occurs first.

3.3.2 Avalanche Breakdown

The collector-to-base junction, when reverse biased, is similar to the diode, and avalanching will cause a large reverse current to flow if the reverse voltage across a junction approaches the value at which the electrons and holes, as they drift in the C–B depletion region, gain sufficient energy to ionize silicon atoms and generate addi-

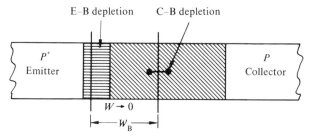

Fig. 3.5 Punch-through of the collector depletion region through the base to the emitter.

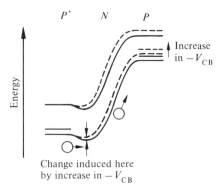

Fig. 3.6 Punch-through energy band diagram for a *pnp* BJT.

tional electron–hole pairs. Under active region operation and a progressively increasing $-V_{CB}$, a point is eventually reached where the collector-base junction begins avalanching (assuming, of course, that punch-through has not occurred). Based on this line of reasoning alone, one would expect output characteristics of the form shown in Fig. 3.7(a) for the common base. The *collector-to-base breakdown voltage, BV_{BC0},* is defined as the B–C voltage where I_C increases very rapidly (typically a factor of 10 to 100 over the low-voltage value) with $I_E = 0$ open circuited. Note that an increase in I_C occurs long before an increase in BV_{BC0}. Instead of thinking of avalanching as occurring at one specific voltage, we must recognize that in reality, a degree of avalanching occurs long before what is normally designated as the breakdown voltage is reached. In fact, some avalanching occurs before any increase in I_C is perceived on the I_C versus $-V_{CB}$ plot.

The common emitter avalanching characteristic is illustrated in Fig. 3.7(b) and is characterized by the *collector-to-emitter breakdown voltage, BV_{EC0}.* First note that BV_{CE0} is much less than BV_{BC0}, and secondly note that the avalanching is more spread out. The breakdown voltage BV_{EC0} is defined with the base open-circuited; that is, with $I_B = 0$. A small amount of avalanching at the C–B junction produces additional electrons that enter the base region, drift through the base (as majority carriers), and supply the additional back injected electrons to the emitter. The additional back injected electrons force a large increase in the number of holes injected from the emitter to the base. These injected holes diffuse to the collector and add a large component to the collector current. Effectively, the current from the avalanching electrons is multiplied by the "transistor action" of the device. Therefore, even at low values of avalanche multiplication, a significant increase in I_C is observed on the output characteristics of the common emitter. Because of the "transistor action" of the electrons reaching the emitter, BV_{EC0} is much less than BV_{BC0}. The reader should note the similarity to the I_{BC0} versus I_{EC0} arguments presented in Chapter 2.

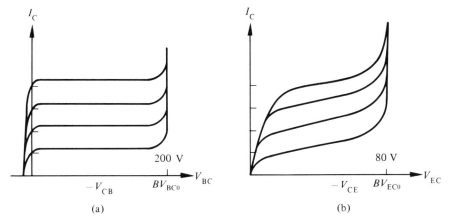

Fig. 3.7 *pnp* BJT avalanching breakdown: (a) common base; (b) common emitter.

As a final observation, note that any resistor placed between the base and the emitter will increase the breakdown voltage for the common emitter. The E–B resistor will "steal" some of the avalanching electrons, prohibiting them from reaching the emitter. With fewer electrons for back injection to the emitter, fewer holes are injected into the base and contribute to I_C. The value of R controls the breakdown voltage as indicated by Eq. (3.23):

$$\underset{R=0}{BV_{BC0}} > BV_{ECR} > \underset{R=\infty}{BV_{EC0}} \tag{3.23}$$

Figures 2.13 and 2.14 illustrate the effect of reducing the number of electrons reaching the emitter by draining them out through the base terminal. In our present case the extra electrons are created in the C–B depletion region by avalanching. As R decreases, more electrons are removed from the base region and cannot be used for injection back into the emitter.

3.4 GEOMETRY EFFECTS

In our ideal model of the bipolar transistor it was assumed that the device was one-dimensional. Practical transistors are always three-dimensional. Figure 1.8 clearly shows the two-dimensional cross sections of a discrete planar BJT and an integrated circuit BJT. The three-dimensional nature of a practical transistor necessitates a revision of the ideal transistor model to take into account several factors.

3.4.1 Emitter Area ≠ Collector Area

The emitter area is not equal to the collector area, as assumed in the simple model. This is illustrated in Fig. 3.8(a) using the discrete planar transistor. We have assumed that the area, A, is that of the emitter. Hence, the inverted operation is affected by the collector area being larger, but the forward operation is quite correct.

3.4.2 Bulk and Contact Resistance

The base current must pass through a resistive region as it leaves the "heart" of the transistor (dotted lines in Fig. 3.8b; called the *intrinsic base*) and before it reaches the base terminal. Therefore, the voltage drop across the junction is actually somewhat less than the terminal voltage V_{EB}. This can be quite serious since the E–B junction is forward biased under normal active mode operation and the emitter current is exponentially related to V_{EB}. Also note that the n^+ base contact has a metal-to-semiconductor junction. This junction is usually ohmic (contact resistance) and contributes to the overall base series resistance r_B.

Figure 3.8(a) illustrates the bulk collector resistance, r_C, which can be significant as the collector is always lightly doped. To reduce r_C, the planar device is made on a

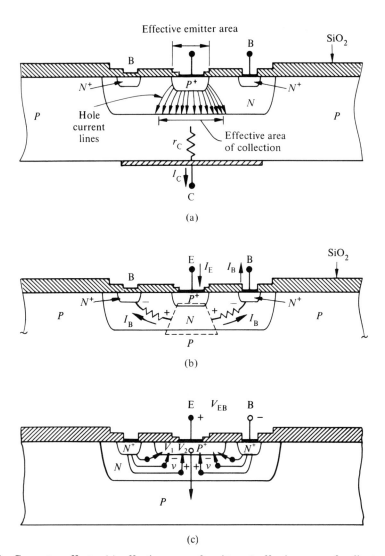

Fig. 3.8 Geometry effects: (a) effective area of emitter $\neq$ effective area of collector; (b) extrinsic base resistance; (c) current crowding due to lateral voltage drops in base; $V_1 > V_2$.

thin (10-μm) epitaxial layer on a p^+ substrate as illustrated in Fig. 1.8(a). The integrated circuit device reduces r_C by placing the "buried layer" of n^+ material below the collector region. The buried layer provides a low-resistance path to the collector contact, as shown in Fig. 1.8(b). Even though the emitters are formed on heavily doped material, the metal-to-semiconductor contact forms a series resistance r_E.

3.4.3 Current Crowding

In addition to a voltage drop in getting to the heart of the transistor, there will also be a voltage drop across the face of the emitter due to the distributed base resistance. The center of the emitter will be less forward biased than the outer edges. Electrons crossing the E–B depletion region from the base (I_{B1}) travel laterally from the base contact as shown in Fig. 3.8(c). Due to the resistance in the base bulk material, the edges have a lower potential because of the IR drop, v. This leads to a larger current around the edges than in the middle of the emitter, a situation known as *current crowding*. Since the periphery of the emitter is more forward biased than the center, the edges reach high-level injection conditions at lower current levels.

The effects of current crowding require a power transistor to have a large emitter perimeter-to-area ratio and many base contacts. Interdigitated emitters and base contacts are often formed into a comb structure by connecting 10 to 20 emitters in parallel with base contacts in between them.

3.5 GENERATION–RECOMBINATION IN THE DEPLETION REGION

The ideal transistor was defined as having no generation or recombination in the E–B and C–B depletion regions. As in the diode, a reverse biased junction with generation in its depletion region will add a generation current component to the reverse saturation current. For example, I_{BC0} will be increased as electrons and holes are generated in the C–B depletion region and fall down the potential hill. Also note that the generation component of the leakage current will increase as $|V_{CB}|^{1/2}$ (for an abrupt C–B junction). The C–B generation current will also increase I_{EC0} due to the additional electrons available for back injection at the E–B junction.

A forward biased junction will have recombination in its depletion region and therefore an additional current component in the emitter and base currents. For a device in the active region, at low values of I_B, Fig. 3.9 illustrates the effect on I_B. Because $I_{En} \cong I_B$ and $I_{En} \ll I_E$, the generation current affects I_B much more than I_E. The arguments for explaining the recombination current in the E–B depletion region are identical to those for the diode. Also plotted in Fig. 3.9 is the collector current, which is not affected by the E–B generation–recombination currents because I_C is primarily due to those holes injected into the base that diffuse to the collector.

The β_{dc} of the transistor is the ratio of I_C to I_B in the active region. By plotting $\ln I_C$ versus V_{EB}, on the same scale as $\ln I_B$ in Fig. 3.9, we find the difference between the plots to be in β_{dc}. In Fig. 3.10, β_{dc} is plotted over many decades of collector current as transcribed from Fig. 3.9. Note that at low current levels the recombination current in the E–B depletion region causes a reduction in β_{dc}. As I_C is increased, the recombination current becomes a smaller part of the total current injected and β_{dc} increases. Beta reaches a peak when the diffusion currents dominate the E–B junction. At very large currents, beta decreases due to the high-level injection and/or the series resistance of the base and collector.

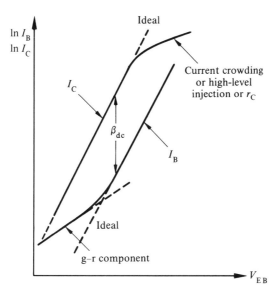

Fig. 3.9 Active region *pnp* with recombination in E–B depletion region and current crowding (or high-level injection).

3.5.1   Current Crowding Revisited

At the larger I_C levels of Fig. 3.9, the effects of current crowding which leads to high-level injection are illustrated. Because the collector current is comprised primarily of the holes injected at the E–B junction, and the E–B is much like the forward biased diode, I_C will not follow the ideal equation forever. At a certain current density level we violate the low-level injection assumption and, as in the diode, the collector current in the active region becomes

$$I_C \cong \frac{qAD_B}{W} e^{qV_{EB}/nkT} \tag{3.24}$$

where $n \to 2$. The condition for high-level injection is aggravated by the "current crowding" at the edges of the emitter; that is, because of the lateral base resistance the edges become more forward biased, leading to the high-level injection condition at lower current levels. The β_{dc} "falloff" with large collector current is illustrated in Fig. 3.10. This effect can also be a result of the base or collector resistance.

3.6 SUMMARY

This chapter has presented several phenomena that explain the deviations of real bipolar transistors from the ideal model. Recombination in the quasi-neutral base region

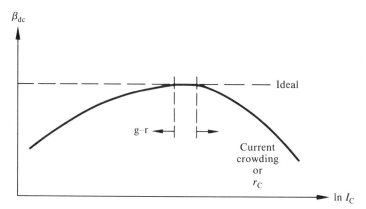

Fig. 3.10   Non-ideal effects on β_{dc}.

W results in a larger base current component, I_{B2}. The Early effect, or base width modulation, explains why the output characteristics have a finite slope as $|V_{CB}|$ is increased. Punch-through and/or avalanching affect the large voltage output characteristics where I_C increases rapidly. The differences between BV_{CB0} and BV_{CE0} were explained in terms of the transistor action of the avalanched C–B carriers. Geometry effects, contact and bulk resistances, and how they affect current flow were discussed. Generation and recombination in the depletion regions were used to explain the β_{dc} versus I_C changes at low levels of current. High-level injection at large values of I_C results in reduction of β_{dc}.

PROBLEMS

3.1 Consider a *pnp* with recombination in the base width W. If W were to be made very large compared with L_B,

(a) Describe what the device would approach.

(b) What will Eqs. (3.8b), (3.9b), and (3.10) become as $W \to \infty$?

3.2 Using the numerical data of $\Delta p_B(0) = 7.883 \times 10^{14}/\text{cm}^3$ and $\Delta p_B(W) = -6.39 \times 10^3/\text{cm}^3$, plot, on the same axis, $\Delta p_B(x)$ for the ideal device and the case of recombination in the base. For the sake of illustration, let $W = 25.4 \ \mu\text{m}$ and $L_B = 46.9 \ \mu\text{m}$. What can be said about the slope of Δp_B at $x = 0$ and $x = W$ as compared with the ideal? What is the area under the two plots?

3.3 Derive an equation for $I_{B1} = I_{En}$ if the emitter width is such that W_E is about equal to L_E. (See Fig. P3.3.)

3.4 Using the data of Problem 2.2 as applied to the *quasi-ideal* device,

(a) Calculate I_{B2} and compare it with I_{B1} and I_{B3}.

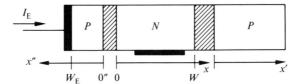

Fig. P3.3

(b) Is base recombination significant?

(c) Calculate α_{dc} and β_{dc} for the quasi-ideal device.

3.5 An ideal $p^{+}np$ device in the active region has $I_E = 961.3$ mA and $W = 3$ μm, when V_{EB} is 0.65 volt.

(a) If base width modulation effects are included, what is the rate of change of I_E with respect to W?

(b) If W is determined only by $W_B - x_n$ where $x_n \cong K(-V_{CB})^{1/2}$, what is the equation for the rate of change of I_E with respect to V_{CB}?

(c) For fixed V_{EB}, determine an equation for the rate at which I_B changes with respect to V_{EC} if quasi-ideal.

3.6 Active region, base width modulation effects provide a slope to the active region $I_C - V_{EC}$ plot with V_{EB} constant. Derive a formula for dI_C/dV_{EC}, assuming an abrupt B–C junction and that $|V_{CB}| \gg V_{bi}$.

3.7 Explain how recombination in the base affects I_{CE0} of the quasi-ideal device versus the ideal pnp.

3.8 Use a sketch similar to Fig. 2.10 for an $n^{+}pn$ device to explain the difference between BV_{CE0} and BV_{CER}.

4 / Small Signal Models

The previous three chapters described the response of the bipolar junction transistor to dc voltages and currents. The present chapter will investigate the response of the transistor to a small signal voltage or current superimposed on the dc values. The term "small signal" implies that the peak values of the signal current and voltage are much smaller than the dc values. Typically, this means a signal voltage of several millivolts or less.

Many small signal circuit models have been developed to represent the signal response. One particularly useful representation is called the *hybrid-pi model*. This model has several advantages that make it particularly attractive to circuit design engineers. For example, the model explicitly relates the signal model circuit element values to the dc operating point variables. Also, the temperature variations in the model parameters are easily obtained, and for frequencies of typically less than 500 MHz the circuit elements are frequency independent.

Small signal models are used for calculating signal gains, input impedances, and output impedances for amplifiers. Because the largest signal gain and least distortion are obtained from bipolar transistors operating in their active region, only active region, small signal models are particularly useful. We assume throughout this chapter that the transistor is operating in the active region.

4.1 LOW-FREQUENCY MODEL

The development of a low-frequency model for the bipolar transistor begins with the assumption of a quasi-static response of the carriers to small changes in the terminal voltages. By "quasi-static" we mean that the electrons and holes return to near steady state in a time much less than the period of the signal. Figure 4.1 illustrates the carrier response to a positive increment of emitter-to-base voltage, ΔV_{EB}, with a period sufficiently long to allow the holes injected from emitter to base and the electrons injected from base to emitter to reach steady state. The figure shows that in this case no signal voltage was applied to the C–B junction. Note that with a larger slope to $p_B(x)$, the emitter current has increased in response to ΔV_{EB}; that is, the total instantaneous value

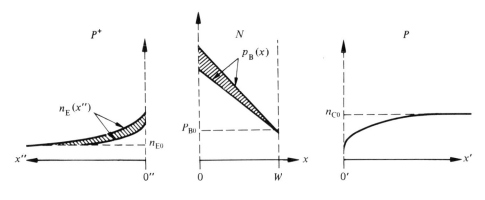

$$p_B(0) = p_{B0}\,e^{\,q(V_{EB} + \Delta V_{EB})/kT}$$

Fig. 4.1 Quasi-static response to V_{eb}.

i_E has increased over the dc value I_E by the signal component, where $\Delta V_{EB} \equiv v_{eb}$ by definition. Equation (4.1) is a formal statement of the emitter current components.

$$i_E = I_E + i_e \tag{4.1}$$

Similarly, the other voltages and currents are written as

$$v_{EB} = V_{EB} + v_{eb} \tag{4.2}$$

$$i_B = I_B + i_b \tag{4.3}$$

$$i_C = I_C + i_c \tag{4.4}$$

$$v_{CB} = V_{CB} + v_{cb} \tag{4.5}$$

Our goal is to develop relationships between i_b and i_c in terms of v_{eb} and v_{ec}. This choice of variables is somewhat arbitrary but is most useful because it models the common emitter amplifier, which has i_b as the input current and i_c as the output current with v_{eb} as the input voltage and v_{ec} as the output voltage.

The Ebers–Moll relationships are a convenient starting point, provided we assume quasi-static responses of the carriers in each bulk region. For low-frequency signal analysis we replace the dc variables in the Ebers–Moll equations [Eqs. (2.45), (2.48), and (2.49)] with the total instantaneous variables of Eqs. (4.1) through (4.5). Before making these substitutions, remember that we are looking for signal models useful for active region operation. Therefore, we simplify the Ebers–Moll equations to active region equations; that is, the E–B is forward biased and the C–B is reverse biased, and $e^{qv_{EB}/kT} \gg 1$ with $e^{qv_{CB}/kT} \ll 1$. Equations (2.45), (2.48), and (2.49) are approximated as

$$i_E = I_{F0}(e^{qv_{EB}/kT} - 1) - \alpha_R I_{R0}(e^{qv_{CB}/kT} - 1) \cong I_{F0}e^{qv_{EB}/kT} \tag{4.6}$$

$$i_C = \alpha_F I_{F0}(e^{qv_{EB}/kT} - 1) - I_{R0}(e^{qv_{CB}/kT} - 1) \cong \alpha_F I_{F0} e^{qv_{EB}/kT} \qquad (4.7)$$

$$i_B = (1 - \alpha_F)I_{F0}(e^{qv_{EB}/kT} - 1) + (1 - \alpha_R)I_{R0}(e^{qv_{CB}/kT} - 1) \cong (1 - \alpha_F)I_{F0} e^{qv_{EB}/kT}$$
$$(4.8)$$

where

$$I_{F0} = qA\left[\frac{D_E n_{E0}}{L_E} + \frac{D_E p_{B0}}{W}\right] \qquad (4.9)$$

and

$$v_{CB} = v_{EB} - v_{EC} \qquad (4.10)$$

The general forms of Eqs. (4.7) and (4.8), upon substituting Eqs. (4.1) through (4.5) and making use of Eq. (4.10) to replace V_{CB} with V_{EC}, are as follows:

$$i_C = f_1(V_{EB} + v_{eb}, V_{EC} + v_{ec}) = I_C + i_c \qquad (4.11)$$

$$i_B = f_2(V_{EB} + v_{eb}, V_{EC} + v_{ec}) = I_B + i_b \qquad (4.12)$$

If v_{eb} and v_{ec} are much smaller than V_{EB} and V_{EC}, respectively, then the terms up to order one in the Taylor formula expansion of Eqs. (4.11) and (4.12) are good approximations to the total instantaneous values. Mathematically the Taylor formula is stated

$$f(x + \Delta x, y + \Delta y) = f(x, y) + \left.\frac{\partial f}{\partial x}\right|_{y,x} \Delta x + \left.\frac{\partial f}{\partial y}\right|_{x,y} \Delta y + \cdots \qquad (4.13)$$

Let's expand Eq. (4.11) for the collector current, i_C, where f_1 is the functional form of Eq. (4.7) with v_{EB} replaced by $V_{EB} + v_{eb}$ and v_{EC} by $V_{EC} + v_{ec}$. In Eq. (4.13) let $x = V_{EB}$, $y = V_{EC}$, $\Delta x = v_{eb}$, and $\Delta y = v_{ec}$, and then we get the following equation for the total collector current:

$$i_C = \underbrace{f_1(V_{EB}, V_{EC})}_{I_C} + \underbrace{\left.\frac{\partial f_1}{\partial V_{EB}}\right|_{V_{EC}, V_{EB}} v_{eb} + \left.\frac{\partial f_1}{\partial V_{EC}}\right|_{V_{EB}, V_{EC}} v_{ec}}_{i_c} \qquad (4.14)$$

We can identify the dc and signal components from Eq. (4.14), of which at this time only the small signal component is of interest. Another assumption is that the partial derivatives in Eq. (4.14) that are evaluated at the total values of the variables are approximately the same as the dc values. Since the signal component is small this is normally an excellent approximation, provided the functions are not too rapidly changing (which they are not).

Consider the last two terms of Eq. (4.14), which are for the signal only and are repeated in Eq. (4.15). In the case of the collector signal current the partial derivatives are current-to-voltage ratios and hence represent a form of conductance. The first term is a ratio of an output variable (i_c) to an input variable (v_{eb}) and is a transconductance which we define as g_m. The second term is a ratio of an output variable to an output variable (v_{ec}) and is a self conductance which is defined as g_0, the output conductance.

$$i_c = \frac{\partial f_1}{\partial V_{EB}}\bigg|_{V_{EC}} v_{eb} + \frac{\partial f_1}{\partial V_{EC}}\bigg|_{V_{EB}} v_{ec} = g_m v_{eb} + g_0 v_{ec} \tag{4.15}$$

Note that the two conductances are defined in terms of the signal short circuits:

$$g_m = \frac{i_c}{v_{eb}} \quad \text{with} \quad v_{ec} = 0$$

and

$$g_0 = \frac{i_c}{v_{ec}} \quad \text{with} \quad v_{eb} = 0$$

The base current is an input variable and is defined as having the functional form of f_2 as applied to Eq. (4.13). Equation (4.16) again shows the form of conductances, and input and reverse transconductance,

$$i_B = f_2(V_{EB}, V_{EC}) + \frac{\partial f_2}{\partial V_{EB}}\bigg|_{V_{EC}} v_{eb} + \frac{\partial f_2}{\partial V_{EC}}\bigg|_{V_{EB}} v_{ec} = I_B + i_b \tag{4.16}$$

where

$$i_b = \frac{\partial f_2}{\partial V_{EB}}\bigg|_{V_{EC}} v_{eb} + \frac{\partial f_2}{\partial V_{EC}}\bigg|_{V_{EB}} v_{ec} = (g_\pi + g_\mu)v_{eb} - g_\mu v_{ec} \tag{4.17}$$

The incremental signal conductances are defined as $g_\pi + g_\mu = i_b/v_{eb}$, the *input* conductance with the output signal $v_{ec} = 0$, and the *reverse feedback conductance* $g_\mu = -i_b/v_{ec}$ with $v_{eb} = 0$.

4.2 LOW-FREQUENCY HYBRID-PI MODEL

The most popular of many possible low-frequency signal models for the bipolar transistor is the hybrid-pi model. Figure 4.2(a) illustrates the simple, ideal, low-frequency device model. If we assume an ideal bipolar transistor with no base width modulation,

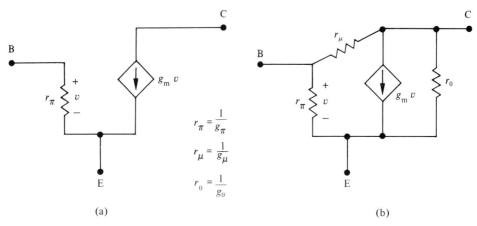

Fig. 4.2 Hybrid-pi models: (a) ideal low-frequency model; (b) nonideal low-frequency model, with base width modulation.

then W is a constant in Eq. (4.9) and I_{F0} is a constant independent of v_{EB} and v_{EC}. Applying Eq. (4.15) to Eq. (4.7) and using Eq. (4.6) for I_E, we find g_m and g_0 as follows:

$$g_m = \left.\frac{\partial I_C}{\partial V_{EB}}\right|_{V_{EC}} = \alpha_F I_{F0}\frac{q}{kT}e^{qV_{EB}/kT} = \alpha_F \frac{q}{kT}I_E = \frac{q}{kT}I_C$$

$$\boxed{g_m = \frac{qI_C}{kT}} \tag{4.18}$$

$$g_0 = \left.\frac{\partial I_C}{\partial V_{EC}}\right|_{V_{EB}} = 0 \tag{4.19}$$

Note the absence of g_0 from Fig. 4.2(a). A similar application of Eq. (4.17) to Eq. (4.8) yields g_π and g_μ:

$$g_\pi = \frac{1}{r_\pi} = \left.\frac{\partial I_B}{\partial V_{EB}}\right|_{V_{EC}} = (1 - \alpha_F)I_{F0}\frac{q}{kT}e^{qV_{EB}/kT} = \frac{q(1 - \alpha_F)}{kT}I_E \tag{4.20}$$

$$r_\pi = \frac{kT}{q(1 - \alpha_F)I_E} = \frac{kT\alpha_F}{q(1 - \alpha_F)I_E} = \frac{kT\beta_F}{qI_C} = \frac{\beta_F}{g_m}$$

$$r_\pi = \frac{\beta_F}{g_m}$$

(4.21)

where

$$\beta_F = \frac{\alpha_F}{1 - \alpha_F}$$

(4.22)

Note that

$$g_\mu = \left. \frac{\partial I_B}{\partial V_{EC}} \right|_{V_{EB}} = 0$$

Note that this model has a perfect (ideal) signal transmission path from input to output with no reverse feedback; that is, a unilateral signal path. The parameter β_F is approximately the same as β_{ac} in the active region and is sometimes written as β_0, the *low-frequency signal beta* of the device.

4.2.1 Nonideal

A bipolar transistor that has base width modulation will have the small signal equivalent circuit of Fig. 4.2(b). In this case W is a function of v_{EB} and v_{EC} (by way of v_{CB}), and g_μ and g_0 are not zero. In most practical cases g_μ is at least 100 times smaller than g_0 and is often ignored in some circuit applications.

4.3 HIGH-FREQUENCY HYBRID-PI MODEL

The high-frequency model for a bipolar transistor can be developed along the lines established for the *p-n* junction since the C–B is a reverse biased diode and the E–B is a forward biased diode. We would expect the C–B junction to have a depletion capacitance that would vary with applied voltage in the same way as a reverse biased diode. Figure 4.3 illustrates the high-frequency hybrid-pi model with a depletion capacitance $C_{jC} = C_\mu$ between the collector and intrinsic base B'. The voltage dependence is that of a reverse biased C–B diode:

$$C_\mu = \frac{C_{\mu 0}}{\left[1 - \dfrac{V_{CB}}{V_{bic}} \right]^m} = C_{jC}$$

(4.23)

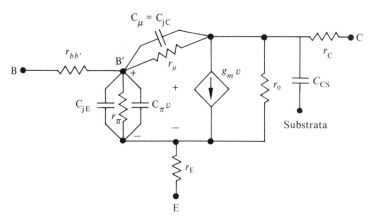

Fig. 4.3   High-frequency hybrid-pi model.

where

$C_{\mu 0}$ is zero biased capacitance, similar to C_{J0} for the *p-n* junction;

V_{bic} is the built-in potential of the B–C junction, identical to V_{bi} for the diode; and

$m = \frac{1}{2}$ to $\frac{1}{3}$ depending on the impurity grading of the junction ($\frac{1}{3}$ for linear, $\frac{1}{2}$ for uniform doping).

The E–B junction has a similar depletion capacitance C_{jE}, with definitions similar to those of Eq. (4.23):

$$C_{jE} = \frac{C_{JE0}}{\left[1 - \dfrac{V_{EB}}{V_{bie}}\right]^{m}} \tag{4.24}$$

where

C_{JE0} is zero biased E–B depletion capacitance;

V_{bie} is the built-in potential of the E–B junction; and

$m = \frac{1}{2}$ for uniform doping and $\frac{1}{3}$ for linear.

The extrinsic base resistance, $r_{bb'}$, of Fig. 4.3 has been discussed previously (in Chapter 3 as r_B) and is primarily the metal-semiconductor contact resistance plus any bulk base region resistance. Resistor r_C in Fig. 4.3 has bulk and contact resistances similar to those of the collector region. A well designed device minimizes $r_{bb'}$, and r_C. In IC devices the collector-to-substrate capacitance is also present and is the depletion capacitance with a dependence similar to Eqs. (4.23) and (4.24).

The emitter base junction is a forward biased junction and will have, in addition to C_{jE}, a diffusion capacitance similar to that of the forward biased diode. In the derivation of the diffusion capacitance of the diode C_D, the *n*-region was considered to have a length much greater than a minority carrier diffusion length. For the *pnp* transistor the base region is much less than a diffusion length and we must modify our capacitance accordingly. A p^{+}-n diode at frequencies where $\omega\tau_p < 1/2$ has

$$C_{\mathrm{D}} \cong \frac{G_0\tau_p}{2} = \frac{qI}{kT}\frac{\tau_p}{2}$$

The minority carrier lifetime is the average time a minority carrier remains alive before recombining. In a transistor we modify this to be the average time it takes a minority carrier to traverse the base width W on its way to the collector, called the *base transit time*, τ_t. The E–B diffusion capacitance is then

$$\boxed{C_\pi = \frac{qI_{\mathrm{E}}}{kT}\tau_{\mathrm{t}} \cong g_{\mathrm{m}}\tau_{\mathrm{t}}} \qquad (4.25)$$

In most cases, where the E–B is reasonably forward biased C_π is much larger than C_{jE}.

The complete hybrid-pi model of Fig. 4.3 is more than adequate for most devices up to frequencies of several hundred megahertz. At larger frequencies the model does not have enough phase shift between the base and collector currents. This is mainly due to our assumption of $\omega\tau < 1$. We encounter the same problem as with the diode; that is, the admittance is complex and C_π varies with frequency. The base region should be modeled similar to a lossey transmission line and will result in an additional phase shift to i_c. One might attempt to model C_π with a frequency dependent capacitor; however, the circuit analysis of such a device becomes a nightmare. For additional models, see the list of suggested readings (Appendix C) near the end of this book.

4.4 SUMMARY

The most commonly used small signal model for a bipolar transistor is the hybrid-pi model, which relates the input signals i_b and v_{eb} to the output signals i_c and v_{ec}. By assuming that the carriers behave quasi-statically, the low-frequency model can be derived with the aid of a Taylor formula expansion of the Ebers–Moll equations. Active region operation models, which are the most useful, allowed us to simplify the Ebers–Moll equations and avoid generating complex mathematical equations. The ideal low-frequency model assumed no base width modulation; hence W and I_{F0} are constants independent of the junction voltages. The ideal model was described with only two circuit elements, a resistance r_π and a transconductance g_{m}.

The low-frequency model with base width modulation has two additional elements, namely g_0 and g_μ, the output conductance and feedback conductance, respectively. Both of these elements represent deviations from the ideal.

A high-frequency model was obtained by extending the low-frequency model to contain capacitances. The C–B was modeled with a depletion capacitance, the E–B with a depletion and diffusion capacitance. Two bulk and contact resistances were added in series with the base and collector to complete the model. For frequencies typically less than several hundred megahertz the hybrid-pi model is an excellent representation of the real device.

PROBLEMS

4.1 For the low-frequency p^+np BJT:

(a) Determine the hybrid-pi model for an ideal BJT if $I_C = 1$ mA, $\beta_F = 200$, and $kT = 0.026$ eV.

(b) Sketch g_m versus I_C for ideal and nonideal cases on the same axis.

(c) Repeat part (b) for r_π versus I_C.

4.2 The p^+np BJT has base width modulation in which the quasi-neutral base width is primarily affected by V_{CB}. If $W = W_{BB} - K_1[|V_{CB}|]^{1/2}$:

(a) Derive a set of equations for the low-frequency hybrid-pi model parameters g_m and g_0.

(b) Discuss how increasing the base doping will affect the terms in the model.

4.3 Apply Eq. (4.13) to the function $i = 2x^2 + 3xy^3 + 3$:

(a) If $x = 1.0 + 0.01$ and $y = 2 + 0.02$, calculate the exact value of i and compare it with the Taylor formula result. What is the percentage error?

(b) In part (a), evaluate the partial derivatives exactly as $x = 1.01$ and $y = 2.02$ and compare them with the approximations of $x = 1$ and $y = 2$.

4.4 For an n^+pn BJT that has $\beta_F = 150$, $C_{\mu 0} = 2$ pF, $C_{JE0} = 10$ pF, $V_{bic} = 0.7$ volts, $V_{bie} = 0.9$ volts, $m = 1/2$, and $\tau_t = 10^{-8}$ s:

(a) Determine the high-frequency model for $I_E = 1$ mA, $V_{BE} = 0.35$, and $V_{BC} = -3$ volts.

(b) Repeat part (a) with $I_E = 0.1$ mA and $V_{BC} = -5$ volts.

4.5 A real measure of the high-frequency BJT is the frequency at which the magnitude of the short circuit current is unity, called f_T. Using Fig. 4.3:

(a) Derive an equation for i_c/i_b with $v_{ec} = 0$

(b) Let r_C, r_E, C_{jE}, C_π, and $r_{bb'}$ all be zero. Determine the frequency at which $|i_c/i_b| = 1$; i.e., f_T.

(c) Use a circuit simulation computer program and calculate f_T for part (a) of Problem 4.4.

4.6 Use the results of Volume II to derive an equation for the base width in terms of the E–B and C–B voltages; assume uniform doping in all regions of the BJT.

5 / Switching Transients

Digital electronics requires that the transistor be switched rapidly from cutoff, through the active region, and into saturation. After a time lapse, the transistor is switched from saturation and returned to the cutoff region of operation. The device physics involved in such large signal transients affects the speed of switching and therefore the circuit design of bipolar logic circuits. To be more specific, the speed at which a logic element can be clocked is determined by how fast the device can be made to switch and by the length of any time delays in the signal path. The goals for this chapter are to develop a model for determining the minority carrier charge storage applicable to large signal switching and to derive the terminal currents as functions of time for the turn-on and turn-off transients.

5.1 CHARGE CONTROL MODEL

The derivation of the minority carrier concentrations for a *pnp* device, as presented in Chapter 2, results in $p_B(x)$, as illustrated by Fig. 5.1(a) for the cutoff, saturation, and active regions. In order to simplify the derivation and concentrate on the most important concepts of switching the device, we assume a p^+np^+ device. By using a heavily doped emitter and collector we force the base minority carrier concentration to be much larger than the minority carrier electrons in the p^+-bulk regions, as shown in Fig. 5.1(b) for the case of saturation. Therefore, when switching from saturation to cutoff, or vice versa, the largest charge component that must be changed is the holes stored in the base region, as the cross-hatched areas of Fig. 5.1(b) indicate. We assume the p^+np^+ bipolar transistor to be approximated by the hole concentration illustrated in Fig. 5.1(c). Note that because $\Delta p_B(x)$ is approximately a straight line, the injected hole concentration can be broken into two parts. The first part, labeled Q_N, is the total injected charge in the base when $V_{CB} = 0$ and $V_{EB} > 0$. The second is Q_I, the total injected charge in the base when $V_{CB} > 0$ and $V_{EB} = 0$. If the total hole charge in the base is Q_B, then

$$Q_N = qA \frac{W}{2} \Delta p_B(0) \tag{5.1}$$

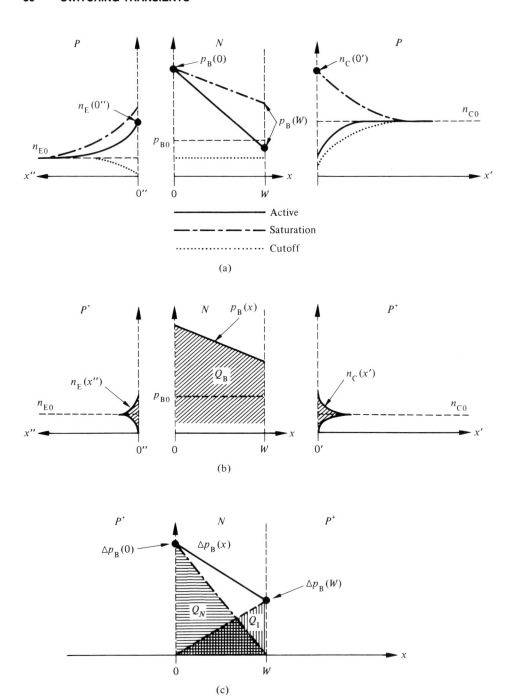

Fig. 5.1 (a) Minority carrier concentration *pnp*; (b) p^+np^+ charge removal in switching from the saturation to the cutoff region; (c) base charge components in the saturation region.

$$Q_I = qA \frac{W}{2} \Delta p_B(W) \tag{5.2}$$

$$Q_B = Q_N + Q_I \tag{5.3}$$

Remember that the boundary conditions at the edges of the base bulk region are $\Delta p_B(0) = p_{B0}(e^{qV_{EB}/kT} - 1)$ and $\Delta p_B(W) = p_{B0}(e^{qV_{CB}/kT} - 1)$.

For the p^+np^+ device in saturation, the emitter current is primarily determined by the slope of $\Delta p_B(x)$ at $x = 0$, and the collector current by the slope at $x = W$, because the electron components are very small due to the p^+-n and n-p^+ junctions. The base currents due to the back injected electrons from the base to the p^+ regions are also small, and we will neglect them. Therefore, base current is primarily that due to recombination of holes in the base; that is, $I_B \cong Q_B/\tau_B$.

A common emitter amplifier has $i_B(t)$ as the input variable that, in our case, controls the base charge storage. The rate of change of $Q_B(t)$ at any instant of time is determined by $i_B(t)$ adding charge to the base region and by recombination removing charge. Arguments similar to those for the p^+-n diode (see Section 6.2, Vol. II) indicate the base charge to be determined by

$$\boxed{\frac{dQ_B(t)}{dt} = i_B(t) - \frac{Q_B(t)}{\tau_B}} \tag{5.4}$$

The base charge is changed by diffusion current and recombination.

The collector current, $i_C(t)$, is the output variable of a common emitter connected bipolar transistor. In the case of active region operation, $\Delta p_B(W) \cong 0$ and $Q_B \cong Q_N$. The collector current can be determined by the total charge in the base that must be transferred to the collector every τ_t seconds, where τ_t is defined as the *base transit time*.* The collector current is

$$\boxed{i_C(t) = \frac{Q_N(t)}{\tau_t} = \frac{Q_B(t)}{\tau_t}} \quad \text{(active region)} \tag{5.5}$$

Inspection of Fig. 5.1(c) for a p^+np^+ device at the edge of saturation, where $V_{CB} = 0$ and $V_{EB} > 0$, is the basis for defining the *base charge at the edge of saturation, Q_{sat}*, as

---

*The transit time, τ_t, is also the average time that it takes a minority carrier to traverse the base region.

$$Q_{sat} = Q_N = I_{Csat}\tau_t \qquad \text{(edge of saturation)} \tag{5.6}$$

where any $V_{CB} > 0$ requires $Q_B > Q_{sat}$. Note that a larger I_{Csat} requires a larger $Q_B = Q_{sat}$.

5.2 TURN-ON TRANSIENT

The switching of a transistor from cutoff to saturation is called the *turn-on transient*. Typically this is accomplished by using a circuit similar to that shown in Fig. 5.2(a). With V_S at zero or some negative value, the transistor is "off," or in the cutoff region of operation, and $v_{EC} \cong V_{CC}$ with $i_C \cong 0$ as indicated by point A in Fig. 5.2(b). When V_S is pulsed positive, the base current increases and v_{EC} decreases toward point B, from cutoff through the active region and into saturation.

Consider the case where $V_S \gg v_{EB}$; then $i_B \cong V_S/R_S = I_B$. That is, the base current is a constant. Applying Eq. (5.4) yields

$$\frac{dQ_B(t)}{dt} = I_B - \frac{Q_B(t)}{\tau_B} \tag{5.7}$$

and if $Q_B = 0$ before the base current is switched,* the solution of Eq. (5.7) results in

$$Q_B(t) = I_B\tau_B(1 - e^{-t/\tau_B}) \tag{5.8}$$

A plot of $Q_B(t)$ is shown in Fig. 5.2(c). Note that for $Q_B < Q_{sat}$, the transistor is switching through the active region on its way to saturation and $Q_B(t) \cong Q_N(t)$. The collector current, $i_C(t)$, is obtained from Eq. (5.5) as

$$i_C(t) = \frac{Q_B(t)}{\tau_t} = \frac{I_B\tau_B}{\tau_t}(1 - e^{-t/\tau_B}), \qquad Q_B \le Q_{sat} \tag{5.9}$$

For most devices τ_B is much larger than τ_t. In fact, as $t \to \infty$, $i_C \to I_C$, and from Eq. (5.9), $I_C/I_B = \tau_B/\tau_t = \beta_F$, provided $Q_B(\infty)$ does not reach Q_{sat} and the device remains active. The general case for i_C is plotted in Fig. 5.2(c), where t_r is the time it takes to reach the edge of saturation. We can solve for t_r by equating I_{Csat} from Eq. (5.6) to Eq. (5.9) with $t = t_r$:

$$I_{Csat} = \frac{Q_{sat}}{\tau_t} = \frac{I_B\tau_B}{\tau_t}(1 - e^{-t_r/\tau_B}) \tag{5.10}$$

---

*Either V_{EB} and V_{CB} are zero, or we neglect the depleted charge ($Q_B < 0$) as small compared with the active region Q_B.

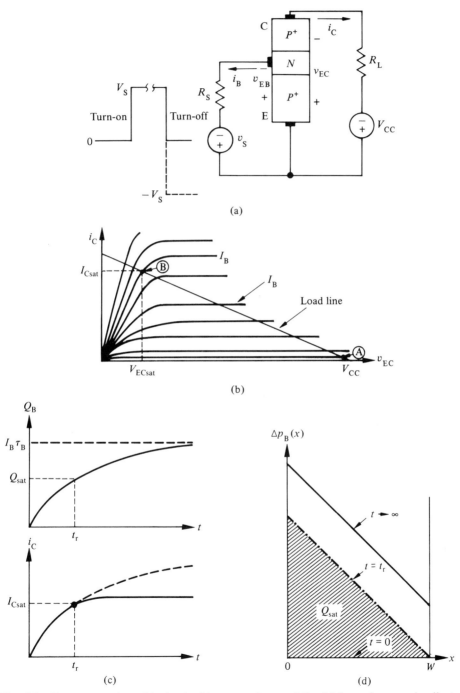

Fig. 5.2 Turn-on transient: (a) circuit; (b) output characteristic; (c) base charge and collector current; (d) injected base charge.

Solving for t_r by taking the natural logarithm yields

$$t_r = \tau_B \ln \left[\frac{1}{1 - \dfrac{I_{Csat} \tau_t}{I_B \tau_B}} \right] \qquad (5.11)$$

where

$$I_{Csat} = \frac{V_{CC} - V_{ECsat}}{R_L} \cong V_{CC}/R_L \qquad (5.12)$$

and

$$I_B = \frac{V_S - v_{EC}}{R_S} \cong V_S/R_S \qquad (5.13)$$

Inspection of Eq. (5.11) indicates that a smaller I_{Csat} and/or a larger I_B will reduce t_r, as will a smaller value of τ_B.

Note in Fig. 5.2(c) that once Q_B reaches Q_{sat}, the collector current increases only a small amount. One way to determine that I_{Csat} is approximately constant is illustrated in Fig. 5.2(d), where the slope of $\Delta p_B(W)$ does not change significantly once in saturation. Another is shown in Fig. 5.2(b), where once in saturation further increases in I_B result in only a slight change in i_C or v_{EC}.

To turn a bipolar logic gate on faster, a larger base current is required. Several logic families, such as Transistor-Transistor Logic (TTL), provide a momentary pulse of current by a capacitive discharge to assist the turn-on transient.

5.3 TURN-OFF TRANSIENT

The switching of a bipolar transistor from the saturation region, through the active region, to cutoff is called the *turn-off transient*. For the common emitter of Fig. 5.1(a), the base current is switched from a value of I_B (in saturation) either to zero or to some negative i_B.

5.3.1 I_B to 0

The base current is switched from I_B, where the device is saturated, to zero as V_S is switched to zero, as illustrated in Fig. 5.3(a). Applying Eq. (5.4), we obtain

$$\frac{dQ_B(t)}{dt} = 0 - \frac{Q_B(t)}{\tau_B} \qquad (5.14)$$

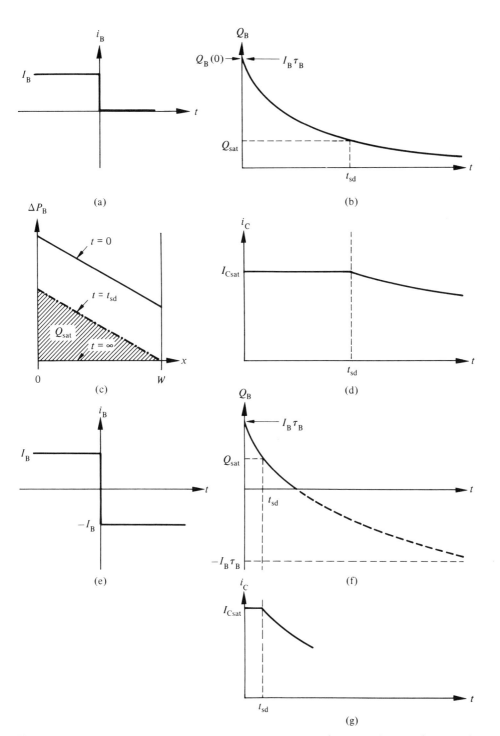

Fig. 5.3 (a) $i_B = 0$ for $t \geq 0$; (b) $Q_B(t)$ for $i_B = 0$; (c) $\Delta p_B(x)$; (d) $i_C(t)$ for $i_B = 0$; (e) $i_B = -I_B$ for $t \geq 0$; (f) $Q_B(t)$ for $i_B = -I_B$; (g) $i_C(t) = -I_B$.

which has a solution

$$Q_B(t) = Q_B(0)e^{-t/\tau_B} \tag{5.15}$$

and is plotted in Fig. 5.3(b). Here $Q_B(0)$ is the total charge under the $t = 0$ line in Fig. 5.3(c), and the further the device is initially into saturation, the larger the value of $Q_B(0)$.

The collector current remains relatively unchanged until $Q_B(t)$ is reduced to Q_{sat}, the time of which is defined as the *storage time delay*, t_{sd}. When $Q_B(t) = Q_{sat}$, the device is at the edge of the saturation and active regions; that is, $Q_B(t_{sd}) = Q_{sat}$. For further time increases the device is in the active region and

$$i_C(t) = \frac{Q_B(t)}{\tau_t} = \frac{Q_B(0)}{\tau_t}e^{-t/\tau_B} \tag{5.16}$$

as plotted in Fig. 5.3(d).

The storage time can be determined from Eq. (5.16) by equating I_{Csat} to $i_C(t_{sd})$:

$$I_{Csat} = \frac{Q_B(0)}{\tau_t}e^{-t_{sd}/\tau_B} = \frac{I_B\tau_B}{\tau_t}e^{-t_{sd}/\tau_B} \tag{5.17}$$

and solving for t_{sd} yields

$$t_{sd} = \tau_B \ln\left[\frac{I_B\tau_B}{I_{Csat}\tau_t}\right] \tag{5.18}$$

Note that a smaller τ_B or I_B will reduce t_{sd}; that is, the less the device is pushed into saturation and/or the smaller the stored base charge, the shorter the storage time delay. Once in the active region, $i_C(t)$ decays exponentially toward zero as shown in Fig. 5.3(d). Because $t_{sd} = 0$ for nonsaturating logic circuits, they are switched faster than saturating circuits. For example, see the Emitter Coupled Logic (ECL) family of circuits, which are the fastest and are nonsaturating.

5.3.2 I_B to $-I_B$

To speed up the removal of minority carrier charge from the base and hence reduce the switching time, the base current is reversed to a value of $-I_B$, as illustrated in Fig. 5.3(e). Equation (5.4) becomes

$$\frac{dQ_B(t)}{dt} = -I_B - \frac{Q_B}{\tau_B} \tag{5.19}$$

which has a solution

$$Q_B(t) = I_B \tau_B (2e^{-t/\tau_B} - 1) \tag{5.20}$$

Equation (5.20) is plotted in Fig. 5.3(f). While in saturation, $Q_B(t) > Q_{sat}$, the collector current is nearly constant, as illustrated in Fig. 5.3(g). When $Q_B(t) = Q_{sat}$, then $i_C(t) = I_{Csat}$ and $t = t_{sd}$.

$$I_{Csat} = \frac{Q_{sat}}{\tau_t} = \frac{I_B \tau_B}{\tau_t} (2e^{-t_{sd}/\tau_B} - 1) \tag{5.21}$$

With some algebraic manipulation, the storage time delay is

$$t_{sd} = \tau_B \ln\left[\frac{I_B \tau_B}{I_{Csat}\tau_t \left[1/2 + 1/2 \dfrac{I_B \tau_B}{I_{Csat}\tau_t} \right]} \right] \tag{5.22}$$

Comparing Eq. (5.22) with Eq. (5.18), we find that t_{sd} has been reduced by using a negative base current ($-I_B$) to aid in removing Q_B from the base region. In Fig. 5.3(f), $Q_B(t)$ is attempting to reach a value of $-I_B\tau_B$; however, our solution ends when $Q_B \cong 0$. Also note that in the active region the collector current is decreasing at a faster rate than in the previous case, when discharging into $I_B = 0$.

In some logic families the base lifetime is purposely reduced by adding impurities with energy levels near the middle of the band gap. The additional recombination helps to switch the device off faster by reducing τ_B. Another IC method is shown in Fig. 5.4. The resistor is used to provide a negative base current to help the turn-off. Note that during the base discharge V_{EB} is of the correct polarity to reverse the base current.

5.4 SUMMARY

The charge control model for the switching of a p^+np^+ device was derived in terms of the base region, minority carrier charge $Q_B(t)$. By assuming p^+ emitter and collector regions, the minority carrier charge storage in their bulk regions is insignificant compared with the base bulk region and the base current can be approximated as the recombination component, Q_B/τ_B. The base charge storage is controlled by the base current adding holes and by recombination removing holes.

A turn-on transient response was modeled, for the common emitter, from cutoff into saturation by solving for the hole charge in the base $Q_B(t)$. At the edge of saturation, $Q_B = Q_{sat}$ and $i_C = I_{Csat}$. Once into saturation, i_C is relatively constant. The turn-on time can be reduced by increasing the base current drive I_B and reducing the base lifetime τ_B.

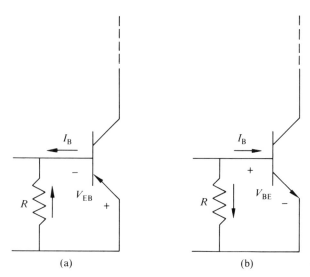

Fig. 5.4 Discharge resistor: (a) *pnp*; (b) *npn*.

The turn-off transient is characterized by two time intervals. One is the storage time, t_{sd}, which is the time required to remove a sufficient amount of base charge to bring the device from deep saturation to the edge of the active region. Once in the active region the collector current can respond to the signal and begins to decrease, which is the second time interval. When switching the transistor with a negative base current, $-I_B$, the charge removal is faster than that when switching with a zero base current; hence there is a shorter switching time. In summary, the turn-off transient is reduced by having less Q_B to change. Therefore, the less the device is pushed into saturation, the shorter the storage time.

PROBLEMS

5.1 If $\tau_B = 0.1$ μsec and $\tau_t = 10^{-9}$ sec, plot t_r/τ_B versus $\ln I_B$ for the case of $I_{Csat} = 10$ mA. Remember that t_r is the time at which the device is saturated. *Hint:* let $100\ \mu A \le I_B \le 100$ mA. What do you conclude about the base drive? Explain why $I_B \ge 100\ \mu A$ is necessary.

5.2 If $\tau_B = 0.1$ μsec and $\tau_t = 10^{-9}$ sec, plot t_{sd}/τ_B versus $\ln I_B$ for $I_{Csat} = 10$ mA (let $100\ \mu A \le I_B \le 100$ mA):

(a) For the case of switching I_B to 0.

(b) For the case of switching I_B to $-I_B$.

5.3 Derive the storage time delay if i_B is switched from I_{B1} to $-I_{B2}$.

5.4 Formulate (do not solve) a method of solution for Fig. 5.4(a) that is required to obtain the discharge time to 10% of the initial base charge as started from the edge of saturation.

Appendix A
Exercise Problems and Solutions

EXERCISE 1.1

(a) What are the E–B and C–B voltage polarities (positive or negative) for a *pnp* BJT operating in the saturation region but having a negative I_C?

(b) Repeat part (a) for the inverted active region.

Solution:

(a) From Fig. 1.3(a) and (b), note that V_{EB} is positive as well as V_{CB} because they are still forward biased to be in saturation. We will learn later that V_{CB} is larger than V_{EB} in this case. We might even call it "inverted saturation" when I_C is negative.

(b) From Fig. 1.3(a) and (b), note that V_{EB} is negative and V_{CB} is positive; i.e., the collector is acting like an emitter and the emitter is acting like a collector.

EXERCISE 1.2

An *npn* BJT has an α_T of 0.998, an emitter injection efficiency of 0.997, and an I_{Cp} of 10 nA:

(a) Calculate α_{dc} and β_{dc} for the device.

(b) If I_B is zero, what is the emitter current?

Solution:

(a)

$$\alpha_{dc} = (0.998)(0.997) = 0.995$$

$$\beta_{dc} = \frac{\alpha_{dc}}{1 - \alpha_{dc}} = \frac{0.995}{1 - 0.995} = 199$$

(b) Since I_{Cp} is 10 nA, then I_{CB0} is 10 nA. I_{CE0} is $(\beta_{dc} + 1)$ times I_{CB0} or

$$I_{CE0} = 200.2 \times 10 \text{ n} = 2002 \text{ n} = 2.002 \ \mu A$$

EXERCISE 2.1

Derive Eq. (2.30) for the collector current starting with the minority carrier diffusion equation.

Solution:

In the p-type collector bulk region, the minority carrier diffusion equation is, from Eq. (2.13),

$$\frac{d^2 \Delta n_C(x')}{dx'^2} = \frac{\Delta n_C(x')}{L_C^2}$$

where $L_C = \sqrt{D_C \tau_C}$.

The solution is given by Eq. (2.7c) as

$$\Delta n_C(x') = C_1 e^{x/L_C} + C_2 e^{-x'/L_C}$$

Applying the boundary conditions at $x' = 0$ and $x' = \infty$ yields

$$\Delta n_C(x' = \infty) = C_1 e^{\infty/L_C} + C_2 e^{-\infty} = 0$$

Therefore, $C_1 = 0$, and

$$\Delta n_C(x' = 0) = 0 e^{0/L_C} + C_2 e^{-0'/L_C} = C_2 = \Delta n_C(0')$$

$$\Delta n_C(0') = n_{C0}(e^{qV_{CB}/kT} - 1)$$

The solution is

$$\Delta n_C(x') = n_{C0}(e^{qV_{CB}/kT} - 1)e^{-x'/L_C}$$

To determine the collector current we apply Eq. (2.4):

$$I_C = -qAD_B \frac{d\Delta p_B}{dx}\bigg|_{x=W} + qAD_C \frac{d\Delta n_C}{dx'}\bigg|_{x'=0}$$

Since the base component is Eq. (2.19), it can be used, and the indicated differentials result in the following after evaluation:

$$I_C = \left[\frac{qAD_B}{W} p_{B0}\right](e^{qV_{EB}/kT} - 1) - qA\left[\frac{D_C n_{C0}}{L_C} + \frac{D_B p_{B0}}{W}\right](e^{qV_{CB}/kT} - 1)$$

EXERCISE 2.2

Derive an equation for I_{BC0} in terms of the Ebers–Moll coefficients.

Solution:

With $I_E = 0$ and the B–C junction reverse biased, $I_C = I_{BC0}$. From Eq. (2.45), with $I_E = 0$,

$$0 = I_F - \alpha_R I_R \cong I_F + \alpha_R I_{R0}$$

since V_{CB} is large and negative, then $I_R \cong -I_{R0}$. Solving for I_F yields

$$I_F = -\alpha_R I_{R0}$$

Then, for $I_E = 0$, substituting into the previous equation yields

$$I_C = I_{BC0} = \alpha_F I_F - I_R = -\alpha_F \alpha_R I_{R0} + I_{R0}$$

and

$$I_{BC0} = (1 - \alpha_F \alpha_R) I_{R0}$$

EXERCISE 2.3

Derive an equation for I_{EC0} in terms of the Ebers–Moll coefficients.

Solution:

By definition, $I_B = 0$ and the device is operated in the active region; that is,

$$I_B = 0 = (1 - \alpha_F) I_F + (1 - \alpha_R) I_R \cong (1 - \alpha_F) I_F - (1 - \alpha_R) I_{R0}$$

since V_{CB} is large and negative, $I_R = -I_{R0}$, and

$$I_C = I_{EC0} \cong \alpha_F I_F + I_{R0}$$

Solving for I_F and substituting it into the above equation yields I_{EC0}:

$$I_{EC0} \cong \frac{\alpha_F (1 - \alpha_R) I_{R0}}{(1 - \alpha_F)} + I_{R0}$$

Further simplification results in

$$I_{ECO} \cong I_{R0} \frac{(1 - \alpha_F \alpha_R)}{(1 - \alpha_F)} = I_{BC0}(\beta_F + 1)$$

This equation clearly illustrates the source of the carriers for I_{ECO} and that I_{ECO} is much larger than I_{BC0}.

EXERCISE 3.1

To verify Eq. (3.5a), let $W/L_B \ll 1$, use the approximation $\sinh(x) = x + \ldots$ for $x \ll 1$, and show that we get Eq. (2.19), the case for no base recombination; i.e., where $W \ll L_B$.

Solution:

From Eq. (3.5a), replace

$$\sinh\left(\frac{W}{L_B}\right) \quad \text{with} \quad \frac{W}{L_B};$$

$$\sinh\left(\frac{x}{L_B}\right) \quad \text{with} \quad \frac{x}{L_B}; \quad \text{and}$$

$$\sinh\left(\frac{W - x}{L_B}\right) \quad \text{with} \quad \frac{W - x}{L_B}$$

The result is

$$\Delta p_B(x) \cong \frac{1}{W/L_B}\left\{[\Delta p_B(0)]\left[\frac{W - x}{L_B}\right] + [\Delta p_B(W)]\left[\frac{x}{L_B}\right]\right\}$$

which, after collecting terms, yields

$$\Delta p_B(x) = -\left[\frac{\Delta p_B(0) - \Delta p_B(W)}{W}\right]x + \Delta p_B(0)$$

EXERCISE 3.2

Show that Eqs. (3.8b) and (3.9b) reduce to the ideal BJT equations of Chapter 2 if $W/L_B <<< 1$.

Solution:

$$\sinh\left(\frac{W}{L_B}\right) \cong \frac{W}{L_B} \quad \text{for} \quad \frac{W}{L_B} \ll 1$$

Therefore, the second bracketed term of Eq. (3.8b) becomes

$$\frac{D_B}{L_B N_B}\left(\frac{L_B}{W}\right) = \frac{D_B}{N_B}\left(\frac{1}{W}\right)$$

Since

$$\coth(x) = \frac{\cosh(x)}{\sinh(x)} \cong \frac{1}{x}$$

then

$$\coth\left(\frac{W}{L_B}\right) = \frac{L_B}{W}$$

and the first bracketed term is

$$\frac{D_B}{L_B N_B}\left(\frac{L_B}{W}\right) = \frac{D_B}{N_B}\left(\frac{1}{W}\right)$$

and the rest of the equation is identical to that of the ideal device.
   Equation (3.9b) follows similarly.

Appendix B
Volume Review Problems and Answers

VOLUME REVIEW PROBLEMS

B.1 List the regions of operation for a bipolar transistor.

B.2 For a *pnp* device, indicate the voltage polarity (+ or −) for the following:

Region	V_{EB}	V_{CB}
Active		
Saturation		
Cutoff		
Inverted active		
Inverted saturation		

B.3 Sketch the thermal equilibrium energy band diagram for a *pnp* transistor.

B.4 Sketch the current components and indicate the electron and hole current flow for an active region (a) *pnp*, (b) *npn*.

B.5 Explain "diode isolation."

B.6 Define the base transport factor for an *npn* device in terms of I_{Cn} and I_{En}. What prevents this from being unity?

B.7 Given that the ideal *pnp* has an emitter injection efficiency of 0.99 and the C–B leakage current is 10 μA, calculate the active region emitter current due to holes if $I_B = 0$.

B.8 Sketch the minority carrier distributions for saturation region operation of a nonideal p^+np.

B.9 For a quasi-ideal *pnp*, sketch the minority carrier distributions for inverted operation.

B.10 Explain the origin of the three base current components for a *pnp* device.

B.11 How are the Ebers–Moll equations modified for an *npn* device? Write the equations and draw an equivalent circuit.

B.12 Why is the emitter doped more heavily than the base?

B.13 Are α_{dc} and γ identical for a quasi-ideal device? Explain.

B.14 Use a carrier flow diagram to explain why I_{CE0} is much different from I_{CB0} for an *npn*.

B.15 "The base region is small" is often stated in device analysis. What is "small" and what is it being compared with?

B.16 Sketch the particle fluxes and their relative sizes for an ideal p^+np device operating in (a) saturation, (b) cutoff, (c) inverted active.

B.17 If W were made smaller, explain how it would affect the base width modulation.

B.18 Why does β_{dc} increase with increasing I_C at small values of collector current?

B.19 Explain why beta "falls off" at large values of collector current.

B.20 Sketch the low-frequency, nonideal, hybrid-pi model. How will the transconductance vary with T?

B.21 Sketch C_μ versus V_{CB} and C_{jE} versus V_{BE}.

B.22 If base width modulation effects were made greater, how would this affect the values of the hybrid-pi model elements?

B.23 Sketch the minority carrier charge storage changes required for a *pnp* device to be switched from cutoff to saturation. Use a cross-hatched area to represent the total change.

B.24 If the base charge is quadrupled, what is the difference between the two storage-time delays when the base current is switched to zero?

B.25 List three methods used to reduce the turn-off time.

ANSWERS TO VOLUME REVIEW PROBLEMS

B.1 Active, saturation, cutoff, inverted

B.2

Region	V_{EB}	V_{CB}
Active	+	−
Saturation	+	+
Cutoff	−	−
Inverted active	−	+
Inverted saturation	+	+

B.3

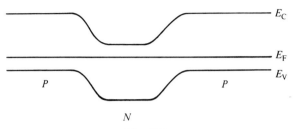

Fig. B.3

B.4

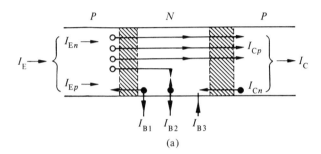

(a)

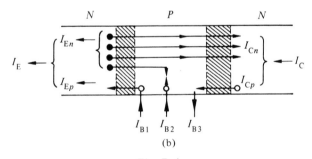

(b)

Fig. B.4

B.5 Diode isolation:

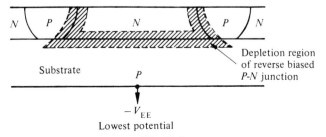

Fig. B.5

B.6 $\alpha_T = I_{Cn}/I_{En}$; any electrons that are recombined in the base.

B.7 $\beta_{dc} = 99$ and $I_{CE0} = 100$ nA

B.8

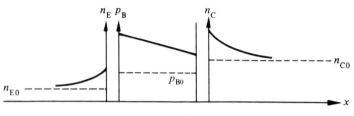

Fig. B.8

B.9

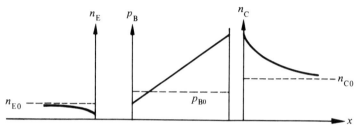

Fig. B.9

B.10 I_{B1} is the back injected electrons from the B–E. I_{B2} is the current supplying electrons for recombination with holes in W. I_{B3} is the electron current from C–B falling down the potential hill; it is thermally generated within one L_C of the depletion region edge.

B.11 p becomes n and n becomes p; V_{BE} becomes V_{EB}; all I's change direction.

$$I_E = I_F - \alpha_R I_R$$

$$I_C = \alpha_F I_F - I_R$$

$$I_B = (1 - \alpha_F)I_F + (1 - \alpha_R)I_R$$

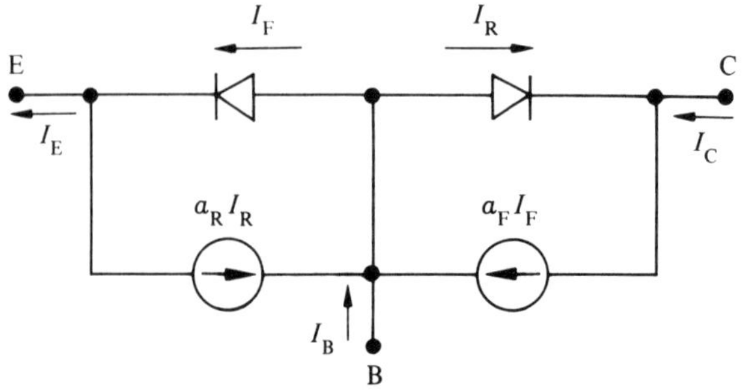

Fig. B.11

B.12 To improve the emitter injection efficiency, and to reduce the back injected carriers from B–E.

B.13 No, because some of the holes are lost to recombination in W; hence I_{Cp} is unequal to I_{Ep} and $\alpha_T < 1$. However, since the recombination current is small compared with I_C and I_E, it is a good approximation that α_{dc} is almost γ.

B.14 $I_{CB0} \approx I_{Cp}$ is formed from thermally generated holes in the collector which provide back injected holes from base to emitter and cause the E–B to be forward biased with a large number of electrons from the emitter (see Fig. B.14.)

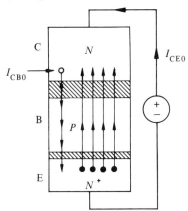

Fig. B.14

B.15 "Small" means $W \ll L_B$, typically $\leq 1 \ \mu$m.

B.16

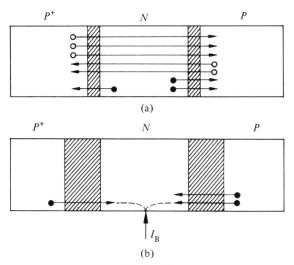

(a)

(b)

Fig. B.16

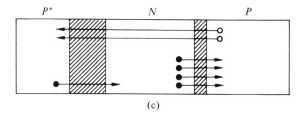

(c)

B.17 ΔW would become a larger percentage of W; therefore it would increase the slope of output characteristics.

B.18 Due to recombination current in the E–B depletion region, which becomes a smaller part of I_B as I_E gets larger.

B.19 At larger I_C, high injection (or r_C) does not increase with the same exponential (r_C effect is not exponential).

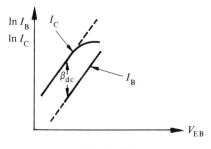

Fig. B.19

B.20 $g_\pi = q(I_C/kT)$; therefore, $g_\mu \propto T^{-1}$ if I_C is constant.

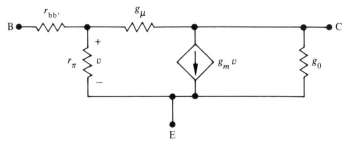

Fig. B.20

B.21

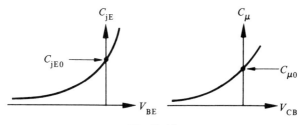

Fig. B.21

B.22 g_μ and g_0 would get larger.

B.23

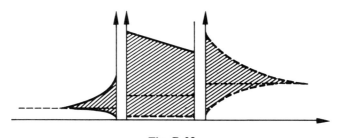

Fig. B.23

B.24 $\tau_B \ln 4$

B.25 Decrease τ_B, decrease I_B, make I_{Csat} larger.

Appendix C
Suggested Readings

A. B. Glaser and G. E. Subak-Sharpe, *Integrated Circuit Engineering*. Reading, MA: Addison-Wesley, 1977. Chapter 2 presents a detailed picture of the junction capacitance, breakdown voltage, and integrated circuit modeling of the bipolar transistor. The base spreading resistance and three-dimensional effects are discussed in detail.

W. H. Hayt, and G. W. Neudeck, *Electronic Circuit Analysis and Design*. Boston: Houghton Mifflin Co., 1976. Chapter 2 is an elementary physical description of the device and its important volt–ampere characteristics. Chapters 3 through 9 are circuit applications.

J. L. Moll, *Physics of Semiconductors*. New York: McGraw-Hill, 1964. Chapter 8 contains a formal discussion of the high-frequency beta and charge storage phenomena.

B. G. Streetman, *Solid State Electronic Devices*. 2nd ed., Englewood Cliffs, NJ: Prentice-Hall, 1980. Chapter 7 is a thorough discussion of the bipolar transistor including the deviations from ideal.

R. S. Muller and T. I. Kamins, *Device Electronics for Integrated Circuits*. 2nd ed., New York: John Wiley & Sons, 1986. An excellent discussion of BJT models.

S. M. Sze, *Semiconductor Devices: Physics and Technology*. New York: John Wiley & Sons, 1985. Has an advanced discussion of nonideal effects.

Appendix D
List of Symbols

BV_{CBO}	breakdown voltage of C–B with $I_E = 0$
BV_{CEO}	breakdown voltage of C–E with $I_B = 0$
BV_{CER}	breakdown voltage of C–E with R from B–E
C_μ	depletion capacitance of C–B junction
$C_{\mu 0}$	depletion capacitance of C–B junction, $V_{\text{BC}} = 0$
C_{jC}	depletion capacitance of C–B junction
C_{jE}	depletion capacitance of E–B junction
C_{JE0}	depletion capacitance of E–B junction, $V_{\text{EB}} = 0$
C_D	diode diffusion capacitance
C_π	E–B diffusion capacitance
D_N	electron diffusion constant
D_P	hole diffusion constant
D_E	emitter minority carrier diffusion constant
D_B	base minority carrier diffusion constant
D_C	collector minority carrier diffusion constant
E_C	conduction band edge energy level (eV)
E_i	intrinsic energy level (eV)
E_F	Fermi energy level (eV)
E_V	valance band edge energy level (eV)
g_m	low-frequency signal transconductance
g_0	low-frequency signal output conductance
g_π	low-frequency signal input conductance
g_μ	low-frequency signal feedback conductance
G_D	low-frequency diode signal conductance
I_B	dc base current

I_C	dc collector current
I_E	dc emitter current
i_B	total instantaneous base current
i_C	total instantaneous collector current
i_E	total instantaneous emitter current
i_b	signal base current
i_c	signal collector current
i_e	signal emitter current
I_N	electron current
I_P	hole current
I_{B1}	base current component due to back-injected carriers
I_{B2}	base current component due to recombination
I_{B3}	base current component due to carriers from the collector
I_{Csat}	collector current at edge of saturation
I_{Ep}	emitter current due to holes
I_{En}	emitter current due to electrons
I_{Cp}	collector current due to holes
I_{Cn}	collector current due to electrons
I_{CB0} or I_{BC0}	C–B saturation current, $I_E = 0$
I_{CE0} or I_{EC0}	C–E saturation current, $I_B = 0$
I_F	Ebers–Moll forward current component
I_{F0}	Ebers–Moll forward coefficient
I_R	Ebers–Moll reverse current component
I_{R0}	Ebers–Moll reverse coefficient
I_S	$= \alpha_F I_{F0} = \alpha_R I_{R0}$
L_N	electron diffusion length (cm)
L_P	hole diffusion length (cm)
L_E	emitter minority carrier diffusion length (cm)
L_B	base minority carrier diffusion length (cm)
L_C	collector minority carrier diffusion length (cm)
m	coefficient of depletion capacitance, $\frac{1}{3} \leq m \leq \frac{1}{2}$
N_A	acceptor impurity concentration ($\#/cm^3$)
n^+	degenerately doped n-type material
N_D	donor impurity concentration ($\#/cm^3$)
N_{AE}	acceptor concentration of the emitter
N_{DB}	donor concentration of the base
N_{AC}	acceptor concentration of the collector

n	ideality factor of junction
n_p	electron concentration in p-material ($\#/\text{cm}^3$)
n_{p0}	electron concentration in p-material at thermal equilibrium ($\#/\text{cm}^3$)
n_{n0}	electron concentration in n-material at thermal equilibrium ($\#/\text{cm}^3$)
n_n	electron concentration in n-material
n_{E0}	electron concentration in emitter at thermal equilibrium
n_{C0}	electron concentration in collector at thermal equilibrium
n_B	electron concentration in base
n_E	electron concentration in emitter
n_C	electron concentration in collector
p^+	degenerately doped p-type material
p_p	hole concentration in p-type material
p_{p0}	hole concentration in p-type material at thermal equilibrium
p_n	hole concentration in n-type material
p_{n0}	hole concentration in n-type material at thermal equilibrium
p_{B0}	hole concentration in base at thermal equilibrium
$p_B(x)$	hole concentration in base
Q_B	minority carrier charge storage in base
Q_N	normal minority carrier charge storage in base, $V_{CB} = 0$, $V_{EB} > 0$
Q_I	inverted minority carrier charge storage in base, $V_{EB} = 0$, $V_{CB} > 0$
Q_{sat}	base minority charge storage at edge of saturation
$r_0 = 1/g_0$	low-frequency signal output resistance
$r_{bb'}$	extrinsic base resistance
r_C	collector bulk and contact resistance
t	time (s)
t_r	rise time of collector current (s)
t_{sd}	storage time delay (s)
V_{EB}	emitter-to-base voltage
V_{CB}	collector-to-base voltage
V_{EC}	emitter-to-collector voltage
v_{EB}	total instantaneous E–B voltage
v_{CB}	total instantaneous C–B voltage
v_{EC}	total instantaneous E–C voltage
v_{eb}	small signal E–B voltage
v_{cb}	small signal C–B voltage
v_{ec}	small signal E–C voltage
V_{bic}	built-in potential of B–C junction

V_{bie}	built-in potential of E–B junction
W	base width (bulk region) (cm)
W_{BB}	metallurgical base width
x, x', x''	x-axis variables
ρ	charge density (coul/cm^3)
$\mathscr{E}$	electric field (V/cm)
α_{dc}	dc alpha
α_T	base transport factor
γ	emitter injection efficiency
β_{dc}	dc beta
β_F	normal forward beta, Ebers–Moll
β_R	reverse beta, Ebers–Moll
ΔV_{EB}	signal component of E–B voltage
Δn_E	injected electron concentration in emitter (#/cm^3)
Δn_C	injected electron concentration in collector (#/cm^3)
Δp_B	injected hole concentration in base (#/cm^3)
τ_E	minority carrier lifetime in emitter (s)
τ_B	minority carrier lifetime in base (s)
τ_C	minority carrier lifetime in collector (s)
α_F	forward alpha, Ebers–Moll
α_R	reverse alpha, alpha, Ebers–Moll
τ_p	hole minority carrier lifetime (s)
τ_t	base transit time (s)
ω	radian frequency

Index

Active region, 4–7, 11–12
Alpha, ideal transistor active region, 17–19, 43
Avalanche breakdown, 66–68

Base charge storage, 86
Base current, 11, 13–14, 30, 34, 35, 57, 65
  components, 11
Base diffusion length, 27
Base minority carrier concentration, 31, 53, 55, 57–58, 86
Base region, 3
Base transit time, 82, 87
Base transport factor, 17, 19
Base width, 62
  metallurgical, 11
Base width modulation, 61–65. *See also* Early effect
  common base, 63
  common emitter, 64
Beta, 18
Beta, ideal transistor, 44, 49
  low-frequency, 80
Boundary conditions, 28
Breakdown voltage. *See* Avalanche breakdown
Built-in potential:
  collector-base junction, 81
  emitter-base junction, 81

Buried layer, 14

CAD, 23, 47
Carrier flux, active region, 16, 36
  cut-off, 39
  inverted, 40
  saturation, 38
Charge control model, 85–88
Circuit symbol, 3–4
Collector contact, 15
Collector current, 29–30, 34, 35, 56–57
Collector diffusion length, 30
Collector reverse saturation current, 44–47
  base open circuit, 45–46
  emitter open circuit, 44–45
  resistance in base, 46–47
Collector region, 3
Collector to base breakdown voltage, 67
Collector to base reverse saturation current, 44
Collector to emitter breakdown voltage, 67
  with resistor, 68
Collector to emitter reverse current, 46–47
Common base, 5, 8
Common collector, 5, 8
Common emitter, 5, 8
  amplifier, 8
  I-V characteristics, 43
Complement of *pnp*, 1, 4–5

115